PRAISE FOR
The Art and Science of Graz

T0177292

"It gives me pleasure to recommend Sarah Flack's *The Art and Science of Grazing*. Sarah offers sound practical information for management of pastures in humid environments. Her years of experience and study allow her to explain the limitations of rotational grazing that were first highlighted by Andre Voisin and to confirm the soundness of Voisin's Rational Grazing."

— Allan Savory, president, Savory Institute

"With feet firmly planted in both practice and science, Sarah Flack introduces pasture-based livestock production in a way that is sure to encourage and empower."

— Joel Salatin, author of *Salad Bar Beef*

"*The Art and Science of Grazing* is a beautiful and highly useful book. Read it, learn how to manage pasture well, then do it! Sarah Flack's good counsel can save your farm and family and bring you happiness. What else is there?"

— Bill Murphy, author of *Greener Pastures on Your Side of the Fence*

"Amidst the flood of information on modern grass farming, Sarah Flack performs nothing short of a rescue operation, distilling the art and science of grazing into an immensely useful book. Experienced hands will learn plenty, but newcomers won't be overwhelmed. *The Art and Science of Grazing* is sure to become one of the most well-worn books on graziers' bookshelves."

— Fred Walters, publisher, Acres U.S.A.

"Sarah Flack has written a wonderful guide to sustainable grazing. She discusses fundamental principles and specific practices that make her book meaningful for anyone interested in the art and science of grazing."

— Fred Provenza, professor emeritus, Department of Wildland Resources, Utah State University

"More and more farmers and ranchers are thirsting for information on animal husbandry that truly protects and regenerates natural resources while enhancing economic viability. Sarah Flack is answering the call. In a readable, user-friendly format, her book provides concrete information and plenty of inspiration. An invaluable resource for anyone raising grazing animals."

— Nicolette Hahn Niman, author of *Defending Beef*

"Sarah Flack's *The Art and Science of Grazing* is an excellent guide for anyone with an interest in the regenerative potential of livestock grazing done right. Whether you are a farmer, rancher, conscientious consumer of meat, or concerned citizen of the planet, Flack's clear, concise prose explains how good grazing is a natural fit in our world."

— Courtney White, author of *Two Percent Solutions for the Planet*

"Intelligently managed grazing has tremendous potential to mitigate climate turmoil, and *The Art and Science of Grazing* contains everything you need to know to do the very best job of grazing. Sarah Flack explores every aspect of holistic pasture management in a most delightful manner that will help all who consult her book, from beginners to the most experienced graziers, to achieve their maximum potential in healing the Earth with livestock and land."

— Jack Lazor, author of *The Organic Grain Grower*

"*The Art and Science of Grazing* is a comprehensive guide for new and experienced graziers. If I had to choose one book as a reference on grazing for both new and experienced livestock graziers, this would be it. The book is well organized, containing many excellent illustrations and beautiful photographs. Sarah Flack shares her knowledge on all aspects of grazing with emphasis on key principles that apply to all farms in humid regions. Multiple case studies provide interest and wisdom gained by farmers who have developed diverse, successful grazing operations throughout the country."

— Guy Jodarski, DVM, CROPP Cooperative/Organic Valley

"I have long been fascinated by the logic of André Voisin's system of Rational Grazing. Sarah Flack's book has persuaded me that this could be a practical option for my own micro-dairy holding."

— Simon Fairlie, author of *Meat*

The Art and Science of
GRAZING

How Grass Farmers Can Create Sustainable Systems
for Healthy Animals and Farm Ecosystems

SARAH FLACK

FOREWORD BY HUBERT J. KARREMAN, VMD

Chelsea Green Publishing
White River Junction, Vermont

Project Manager: Alexander Bullett
Acquisitions Editor: Makenna Goodman
Developmental Editor: Fern Marshall Bradley
Copy Editor: Laura Jorstad
Proofreader: Eric Raetz
Indexer: Linda Hallinger
Designer: Melissa Jacobson

Printed in the United States of America.
First printing June, 2016.
10 9 8 7 6   22 23 24 25 26

Our Commitment to Green Publishing
Chelsea Green sees publishing as a tool for cultural change and ecological stewardship. We strive to align our book manufacturing practices with our editorial mission and to reduce the impact of our business enterprise in the environment. We print our books and catalogs on chlorine-free recycled paper, using vegetable-based inks whenever possible. This book may cost slightly more because it was printed on paper from responsibly managed forests, and we hope you'll agree that it's worth it. *The Art and Science of Grazing* was printed on paper supplied by Versa Press that is certified by the Forest Stewardship Council®.

Library of Congress Cataloging-in-Publication Data
Names: Flack, Sarah, 1969- author.
Title: The art and science of grazing : how grass farmers can create
   sustainable systems for healthy animals and farm ecosystems / Sarah Flack.
Description: White River Junction, Vermont : Chelsea Green Publishing, [2016]
   | Includes bibliographical references and index.
Identifiers: LCCN 2016000954| ISBN 9781603586115 (pbk.) | ISBN 9781603586122 (ebook)
Subjects: LCSH: Livestock systems. | Grazing. | Pastures--Management.
Classification: LCC SF140.L65 .F53 2016 | DDC 636.08/45--dc23
LC record available at http://lccn.loc.gov/2016000954

Chelsea Green Publishing
85 North Main Street, Suite 120
White River Junction, VT 05001
(802) 295-6300
www.chelseagreen.com

CONTENTS

FOREWORD

The world of livestock farming has become split between those farms that exert complete control over animals and environment to force maximal production and those farms that graze and allow animals, plants, and land to interact and thereby attain optimal production. This book is an elegant and useful guide for the latter. Sarah Flack provides all the angles needed to ensure a well-planned and well-managed grazing system.

The need for change in our livestock-rearing practices has never been greater. Billions of dollars have been spent at land grant universities on research to fine tune confinement farming over the last 50 years. The farmer has been removed from the role of steward of the resources on the farm and forced to act merely as a collection vessel for the products of agribusiness interests.

The industrialization of agriculture has culminated in creating genetically modified organisms that farmers are told they can't do without. Agribusiness has thrown the Precautionary Principle out the window into a wind that is already carrying genetically engineered pollen as far as the wind can travel.

Confinement agriculture leaves its waste upon our earth in many ways, from huge manure lagoons to the constant use of antibiotic crutches that have a "trickle up" effect of potential antibiotic resistance upon higher levels of the food chain, notably upon humans. Confinement farming is also heavily dependent upon petroleum, not only for diesel fuel but also as the raw material for its fertilizers, herbicides, insecticides, and fungicides.

In contrast, farms that practice grazing depend less upon prescribed inputs and instead rely upon biological principles that have withstood the test of time, as Sarah Flack explains so well in *The Art and Science of Grazing*. The main source of energy on grazing farms is the sun. When the functional anatomy of plants and animals is allowed to proceed as biologically designed — by the continual interplay of animals grazing plants in a smart manner — the land becomes restored both in structure and fertility while at the same time providing true health to the animals eating fresh feeds.

This book gives clarity to the sometimes seemingly independent factors of soil, plants, and animals in the ecosystem, beginning with soil-plant relationships, then adding in the plant-animal relationships, and finally applying the concepts to ensure a well-designed grazing system to fit any farm, anywhere.

For animals living directly upon well-managed pastures, their food supplies biologically healthy energy, protein, and fiber as well as naturally occurring vitamins and minerals. The bounty of the land is self-perpetuating when animals and plants are managed in a manner similar to how bison once roamed the prairies, never staying in one place too long. The more that farm animals can live like their wild relatives — by moving from paddock to paddock within a farm's perimeter — the less likely they are to suffer from illness. The result is highly nutritious food for society.

Readers who have not taken agronomy or animal science courses will appreciate Sarah Flack's easy-to-read, logical explanation of functional anatomy and physiology of plants and animals. She concisely covers every point needed to enhance soil health, plant health, and animal health, with scientific references for those wishing to delve further. One of the many positive effects of implementing a successful grazing system is a reduced need for veterinary intervention because the farm animals live as they were designed to: walking through the landscape eating fresh feeds of nutritional and medicinal value with plenty of exercise. As a person who was originally trained in soil science, practiced management-intensive grazing as a herdsman, and then trained in veterinary medicine, I am happy to say that this book covers it all in a friendly, inviting way. I wish I had had this book when I was a herdsman.

Those new to grazing are especially lucky with the information provided herein, for it encompasses a career's worth of the author's direct observations, practical experience, and formal education. This book will provide a novice grazier the knowledge of what to expect when embarking on pasturing animals, as well as guidance that will enhance both a farm's biological sustainability and economic profitability. More seasoned graziers will pick up welcome insights and tips that will help them make adjustments in their systems. Especially applicable to those already grazing are the individual farm profiles (chapter sub-sections called "The Art of Good Grazing") that bring to life the concepts and details presented. The appendices provide practical information in a condensed format.

The hallmark of this book is that it provides many options from which to choose that are applicable to any specific geographic location. This is in contrast to the rigidly controlled world of confined animal feeding operations (CAFOs), which are essentially well-fed and well-medicated concentration camps that maintain a repetitive, homogeneous, blueprint design approach that hardly takes into account any local specificity. By contrast, grazing farms can easily adjust to changing conditions just as soil, plants, and animals always have done.

Sarah Flack is in tune with the needs of practical-minded farmers yet also respectful of the biology that underlies the whole system. She always reminds us of the continual, interactive forces that each component of the grazing system exerts on the others. The result is at once both functional and inspiring, just as agro-ecology truly is – a system that creates healthy soil, plants, and animals, as well as profit for farmers while honoring the ecosystem of which we are all part.

HUBERT J. KARREMAN, VMD
Lancaster, Pennsylvania
January, 2016

Transforming the Landscape Through Grazing

When pastures are managed well, the complex interrelationships of the farmer, livestock, plants, and soils within the ecosystem create a beautiful dance and flow. Well-managed grass-based livestock systems are attractive to look at, but—more important—they are beneficial to our environment and well-being. As you learn more about the science of well-managed grazing systems, you'll find the elegance of the pasture ecosystem even more fascinating. Using good grazing practices creates an ecosystem that is more diverse, healthy, and productive. I have watched this positive transformation on many farms over the years, and it always fills me with awe and reverence.

My interest in pastures and livestock started early, and it led me to a lifetime of study and teaching about both the science and the practical everyday management of grazing systems. As a child, I lived with my American/English family in a small farming and fishing community on the South Island of New Zealand. Behind our house, sheep grazed on a hillside pasture most of the time. On the other side of our house was a cow pasture. Even as a young child, I noticed the difference between the grass in the cow pasture and that in the sheep pasture. The cow pasture had more vegetation and softer soil, so it was nicer to fall off my bike there. The sheep pasture usually had much shorter grass, more weeds, and muddy gullies (in which we kids enjoyed playing).

One very rainy night, a sudden loud noise woke me up. The house was vibrating. I assumed it was just a little earthquake, because these were quite common in New Zealand. The next morning I learned that a large section of the sheep pasture behind the house had slumped. Earth had broken loose from the hillside and hit the back of our home. We were lucky that the damage from this landslide was minimal and none of us were hurt.

Conversations with my parents over the next few days helped me learn more about why this had happened. I learned about erosion from my mom. Over breakfast, she explained "mass wasting" to me, which is when large amounts of earth slump or slough downhill. I learned about ecology from my dad, who told me about how continuous grazing pressure from livestock can damage plants. It was a memorable lesson that poorly managed grazing could do a lot of damage to our ecosystem and even endanger the families living there. Because the sheep pasture was continuously grazed, the plants never had a chance to recover. But the cow pasture was only grazed for a few days at a time, and then the plants were allowed to rest and regrow for several weeks before being grazed again.

As I continued to spend time in the pastures around my house and at friends' dairy farms, I noticed the different livestock grazing systems, the mix of plants, the diversity of soils, and how different the pastures could be from one another.

Over the years I have had the chance to travel to many countries and observe firsthand a wide range of soil types, vegetation, and livestock. I learned to ask questions about the ecosystems: What had the native plant and animal species been? How had the ecosystem changed over time due to human management for food production? So began my lifelong fascination with the relationships among plants, grazing animals, and the humans who make management decisions and create changes — both for the better and worse — in our ecosystems. Most of the regions I have worked in are humid (non-brittle) ecosystems or irrigated pastures. Thus, in

Figure I.1. Erosion due to badly managed grazing and other poor practices occurs in several forms, including surface washing of soil, small channels called rills, and deeper gullies, as seen here.

Figure I.2. Soil erosion caused by poor grazing practices on steeper slopes can result in significant amounts of soil slumping downhill.

this book I don't cover topics related to dryland grazing in non-humid (brittle) ecosystems.

Once my parents and I moved back to the United States, my parents started a grass-based livestock farm in northern Vermont. The soils and climate there were very different from New Zealand. Over the years the family farm included cows, sheep, poultry, and sometimes a few goats. In the early years, frustrated by the lack of suitable electric fence supplies in the area, my parents started a Gallagher high-tensile fence business. I learned more about animal husbandry, fencing, and grazing management as our family used what was then called mob stocking. This method employs high-stock-density grazing to transform a brushy, overgrown series of fields into high-quality pasture. This transformation was done almost entirely with livestock, even though in the early years large areas of the farm were so overgrown there was little or no grass or clover present.

Each area was grazed for a day or less with a high-stock-density group or "mob" of sheep, goats, and sometimes dairy heifers or beef cattle. The animals repeatedly defoliated the brush; within a few years, the woody plants died and higher-quality grass and legume plants filled the pasture.

This vivid transformation taught me how grazing-adapted species of plants thrive under good grazing management, while the "weeds" that are not well adapted to being grazed are removed from the pasture. I also learned that a thoughtfully planned grazing system can significantly improve an ecosystem, in sharp contrast with the damage that poor grazing practices can cause.

I was lucky to have the opportunity to further study both the science of management-intensive grazing, which was developed by grazing consultant Jim Gerrish, and Allan Savory's holistic planned grazing, first as an undergraduate student and then in graduate studies

with Dr. Bill Murphy (author of *Greener Pastures on Your Side of the Fence*) and Dr. Nthoana Mzamane at the University of Vermont. During my graduate studies, I observed how white clover and other pasture plants responded when allowed to grow taller before being grazed. We also studied how they responded when we let the herd graze the plants down severely, compared with leaving more plant material behind after grazing.[1] In the same pasture, other students studied earthworm populations, the behavior of the cows as they grazed, and other pasture-related subjects.

I found the science of the natural and the agricultural ecosystem fascinating. But ultimately, the moment that allowed me to clearly see the importance of this knowledge came during my graduate thesis defense. After I had presented the detailed science of how different grazing systems had created changes in the pasture plants and thus the amount of photosynthetically active light that penetrated the pasture plant canopy, one of my advisers asked me a very simple question: "What does this actually mean for farmers?"

My answer was, and continues to be, that farmers and their farms will benefit greatly from understanding that, in partnership with their animals, they can create profound change in pasture quality and productivity and the

Figure I.3. With good grazing management, a dense healthy community of plants and residue protects soil from erosion. This pasture, which is being measured with a grazing stick, is on my family farm.

Figure I.4. During the early years on my family's farm, the primary plants were small trees and brush species, with few grasses or legumes.

performance of the livestock. Grazing-adapted plants are amazingly responsive to our management decisions. As farmers we have a huge impact on and responsibility for our local farm pasture ecosystems — and also for the larger planetary ecosystem we all share. This profound knowledge empowers farmers.

I first thought through that crucial question over 20 years ago, and I have been farming or serving as a grazing consultant ever since. I continue to be impressed by how good stewardship of land, using livestock, can create many benefits. Farmers can cause the species composition of a pasture to transform, without tillage and reseeding, simply by using grazing animals wisely. Soil health can improve, animal welfare can benefit, feed costs can go down, animal performance can increase, and farm finances can become more sustainable. Each year we also learn more about other benefits from good grazing management, including better human nutrition and improvements to the health of the larger planetary ecosystem we all share.

Good grazing systems allow farmers to create positive change in their landscapes, livestock, and checkbook. As our climate becomes more unpredictable, and the costs of fuel, purchased feed, and other farm inputs rise, having a well-designed and well-managed grazing system is essential. In addition, as we learn more about the human health risks from conventional agriculture's overreliance on synthetic chemical inputs, we can see that proficient grass farming provides an alternative that can reduce or even eliminate use of these risky materials. We need more farms that have a positive effect on the environment and produce healthy food, and that is what well-managed grass-based farms can do.

The purpose of this book is to empower farmers to ask the right questions and to create grazing systems that are truly effective at meeting their farm and quality-of-life goals. A well-designed grazing system will take into account the local soils, climate, landscape, plant species, types of grazing livestock, and needs of the farmer. By thinking about all these important parts of the farm ecosystem, farmers can design the best possible grazing system. Instead of feeling overwhelmed by the complexity of the "parts" that make up the farm ecosystem, a farmer can instead step back and appreciate the elegance of how well these natural systems work together.

When designing a grazing system, though, it's also crucial to keep in mind that all good grazing systems are based on critical underlying guidelines or principles. These basic grazing guidelines have been known for at least 200 years; many scientists and farmers have written about them. I state and restate these principles many times in this book. If you read it from cover to cover, the repetition will help make these principles the guiding force of your work with grazing systems. If, instead, you read only a few chapters or use the book as an occasional reference guide, you'll still most likely encounter a discussion of these highly important guidelines.

Throughout the book you'll find descriptions of grazing systems in use on a variety of working farms. Each of these is a unique grazing system for sheep, beef, goats, or dairy cattle. These real-life profiles can serve as a source of practical ideas, but also inspiration. There are so many creative ways that farmers are applying the basic principles of good grazing management to meet their goals. At the same time, these farmers are able to improve the health of a piece of this precious planet we all share.

PART ONE

Laying the Groundwork

A well-managed grazing system can create incredible benefits for the resilience of our planetary ecosystem, the well-being of livestock, human health, and the financial sustainability of farms. Grazing can be a powerful tool for positive change, but it can also be damaging to the pasture ecosystem when poorly planned and managed. It is important to understand the differences among types of grazing systems, and how different management systems either do or don't function to provide high-quality forage and improve the pasture ecosystem. By developing a thorough knowledge of the basic guidelines of good grazing management, it becomes possible to choose the right grazing system for your own farm.

CHAPTER 1

Benefits of Good Pasture Management

Have you noticed that some pastures seem to run out during the grazing season, leaving you with no good option other than providing more supplemental feed to your animals? Have you ever wondered why some pastures become weedy, stay short, or are full of plants the animals won't eat? Many of these problems are caused by poorly designed grazing systems, or they result from damage to pasture plants and soils caused by poor management.

When done well, grazing management can improve animal well-being, ecological health, and the financial sustainability of the farm. An effectively designed and managed grass-based livestock operation requires that the farmer understand the basic principles of grazing management and ecology. It also requires an understanding of what pasture plants and livestock need, and how to put that information together with the right infrastructure. This knowledge makes it possible for the farmer to choose which type of grazing system best fits the farm and family goals, and customize it so that the system is practical and works well.

The pasture ecosystem includes many interrelated parts including plants, soils, animals, local weather, and the farmer. (Yes, humans are part of the ecosystem!) Many factors affect the types of plants in a pasture and how vigorously they grow. Such factors can include how animals are being used to harvest the plants and how well manure is being distributed, as well as the impact of hooves, the weather, and the past and current management of soil fertility and health. Grazing systems that are designed and managed with an understanding of these factors create many benefits. However, poorly designed and managed pastures can lead to many problems, including negative impacts on the ecosystem.

For beginner farmers or those new to grass-based livestock farming, the number of suggestions on the "best way" to improve soil fertility, forage quality, and pasture production can be overwhelming. This is particularly challenging for farmers who have not yet learned the fundamental principles of good grazing management. Without a solid understanding of the basic guidelines of how to set up and manage a pasture system, it's easy to get sidetracked by the latest fads and spend money on unnecessary or impractical infrastructure and inputs. This can result in a system that doesn't meet your quality-of-life goals or the financial needs of the farm, and it can jeopardize the productivity and welfare of the livestock.

To be a good grass farmer, you need technical and scientific knowledge of grazing management and animal husbandry, plus the observation and monitoring skills to see subtle changes over time in livestock, soils, and pasture plants. With this knowledge, farmers have found diverse creative ways to apply good grazing principles to many types of land bases with different types of livestock.

How Grazing Improves Pasture

When done correctly, livestock grazing can create many benefits for the environment, plants, soils, animals, and farm income. Good grazing management can cause the mix of plant species in a pasture to change even without tillage and reseeding, simply as a result of animal impact. When animals spend time in a pasture, they do more than just eat. They trample weeds and dead plants into the soil, which adds organic matter and can improve soil biological activity. They choose certain plants to eat, and they spread their own manure. In just a few years, this combination of activities can convert weedy brushy pastures, where animals have to search to find good-quality forage, into highly productive pastures capable of supporting more animals and providing a higher quality of forage. As pastures improve, plant density and diversity increase, which protects soils from erosion and compaction. In addition to the visible aboveground improvement, there is also increased plant root growth and better cycling of nutrients through the soil so they are more available to plants and other soil life.

Well-managed pasture provides low-cost, high-quality feed. This is particularly helpful for farmers who must deal with rising costs of purchased feed, combined with the expense of harvest and storage of forages. With good pasture management, most sheep and beef farms should be able to completely eliminate feeding grain or stored forages during the grazing season. For some dairy farms, it may be possible to significantly reduce or, in some cases, completely eliminate grain feeding. However, at the time of this writing it is probably still more profitable for most dairy farms to supplement pasture with grain, due to the current cost of grain and pay price for milk. (See chapter 11 for more detail on 100 percent grass feeding.)

In addition to lower costs due to reduced supplemental feed purchases, there are also fewer labor and input costs because the livestock harvest their own feed, which reduces or eliminates the need for mechanical harvest, storage, and feeding. There will still be some expense when starting a grazing system, because fencing and some investment in seed and soil fertility may be needed. Still, the benefits of a well-managed system will generally cover the costs!

When pastures are designed and managed well, there will be a better seasonal availability of high-quality pasture. This can allow grazing to start earlier and run later into the fall or winter. This lowers the expense of purchased fuel, labor, and equipment repairs. For example, a small beef herd with just 20 head can consume 8 to 10 round bales a week. At $40 per bale, the farmer can save as much as $400 for each week the grazing season is

BENEFITS OF GOOD GRAZING MANAGEMENT

Improved financial sustainability is an important benefit of well-designed and -managed pastures. But lower costs and higher yields are just two of the many positive impacts from grass-based agriculture. There will also be improved living conditions and animal well-being, along with a whole array of ecological benefits.

Benefits to Farmers

- Lower feed costs.
- Improved pasture yields.
- Higher forage quality.
- Healthier, longer-lived animals (which results in additional income from the sale of livestock and products).
- Less money and time spent on weed control, forage harvesting, and manure spreading.
- Less need for expense and labor related to reseeding.
- New market opportunities to health-minded consumers.

Benefits to Animals

- Improved livestock health and welfare.
- Improved quality of life.
- Freedom to engage in natural behaviors with the rest of the herd or flock.

Benefits to the Planet and Humans

- Increased levels of healthy nutrients in grass-fed meat and milk.
- Less exposure to pesticides and chemicals.
- Improved environmental and ecosystem health.
- Healthier soils and plant communities with vegetative cover (which improves water infiltration and quality, and reduces erosion).
- Carbon retention in plants and soil instead of in the atmosphere (carbon sequestration).
- Energy savings from less use of fossil fuel.
- Reduced or eliminated use of herbicides.
- Potentially slower spread of antibiotic-resistant pathogens (due to reduced use of antibiotics).

MESIC VERSUS DRYLAND

The types of grazing systems discussed in this book are best suited for mesic (humid) regions. This book doesn't cover the topic of dryland or brittle ecosystem grazing, which requires a different approach.

extended into the fall. This savings can be reinvested in more fence, seed, or fertility inputs. Or it can go into the vacation savings account!

Benefits to Livestock and Humans

Good grazing management can improve livestock health due to better nutrition, lower stress, and greater opportunity for natural herd or flock behavior. This can reduce cull rates, produce longer-lived animals, lower vet bills, and make it possible to earn additional income from sales of livestock. Cattle on pasture also spread their own manure and mow the weeds.

When their diet is mostly or entirely from pasture, animals produce meat and milk with different amounts and types of nutrients than what grain-fed livestock produce. The nutrients that appear in higher amounts in grassfed meat and milk include carotenoids, vitamin E, omega-3 fatty acids, conjugated linoleic acid (CLA), and other nutrients.

Research on the human benefits of increasing dietary CLA intake indicates that doing so may lower the risk of cancer and heart disease. Research also shows that increasing omega-3 fatty acids in relation to omega-6 fatty acids may reduce the risk of cancer, obesity, diabetes, and other illnesses.[1] Grassfed products are also higher in healthy antioxidants (including vitamin E) and carotenoids such as lutein, zeaxanthin, and beta-carotene.[2]

This information can be used to provide some new marketing opportunities. It is not surprising that the higher nutritional value of these foods attracts health-minded consumers, and many consumers also appreciate

Figure 1.1. The lambs in this high-quality pasture are grazing a diverse mix of plants including red clover, dandelion, plantain, white clover, and several cool-season perennial grasses.

the other benefits of grass farming, including improved animal welfare and environmental health. Many of the ecological benefits of well-managed pasture-based production systems are a result of having higher-density populations of deep-rooted perennial plants and healthy, biologically active soils. Benefits from the perennial soil cover include decreased soil erosion and improved water quality due to improved soil stability and less nutrient runoff. In addition, with the livestock doing the harvesting and manure spreading, there is less fossil fuel used and fewer emissions of greenhouse gases, and carbon is actually sequestered in plants and soil. These perennial pastures are highly diverse, which also improves wildlife habitat.[3]

The Art of Good Grazing
Healing Pastures Through Good Grazing

When Brooke Henley and Tom Garnett started managing their Maryland farm in 2012, some fields were completely bare and lifeless due to previous cropping with sorghum and corn. Overgrazing had damaged other areas, which were a mix of stunted plants, weeds, and bare soil. But Brooke and Tom have transformed the fields of Spring Pastures Farm into productive pastures with a growing herd of beef cattle in less than three years of good management.

Brooke and Tom did their research on what plant species would do well on their farm, and they studied the science of good grazing management and soil fertility. They reseeded the bare fields with a pasture mix in 2013, and used good grazing techniques with the beef herd to improve all of the fields. The plants in the pasture mix didn't all survive, and they had to reseed a few areas. Overall, though, there was enough diversity of species in the pasture mix for success. Brooke and Tom say they still have more work to do improving the land, but their progress in less than three years is an impressive demonstration of the healing and regeneration that good grazing practices can bring to the land.

During cooler spring and fall weather, the pastures provide the beef herd with a variety of cool-season perennial grasses and legumes including orchard grass and white clover. In midsummer, Johnsongrass still pops up in a few areas on the farm. This warm-season grass, which grows well in midsummer, is considered a weed by many farmers. However, Tom and Brooke find that if managed well, Johnsongrass provides reasonably good midsummer forage for the herd. Some fields are planted with tall fescue. This plant also grows well in hot weather, but they always plan to take the animals off those fields by mid-August. This provides the perfect timing to let the tall fescue grow, providing stockpiled forage for fall and winter grazing. The herd loves to graze the tall fescue in January, and this stockpiled pasture reduces the amount of hay they need to be fed. (See chapter 6 for more detail on Johnsongrass and tall fescue.)

Figure 1.2. This photo from 2012 shows the poor state of some areas of the farm when Brooke Henley and Tom Garnett first stepped in as managers. Photo courtesy of Brooke Henley, Spring Pastures Farm.

Brooke and Tom plan to use short-duration grazing with long recovery periods to keep progressing toward their goal of a high diversity of perennial species. They move the herd to new pasture at least once a day; pastures are then left to regrow for three to five weeks. Tom walks the pastures every morning to see how the plants are growing and how the herd is doing. Their daily record of how long the herd stays in each area provides helpful information as they plan future grazing rotations and slowly grow the herd size.

At this point they don't plan to do any additional seeding, since plant density and quality are continuing to improve through good grazing practices alone. They are pleased with the forage quality and the steady improvement in performance of the beef herd.

The farm is about half forest and half open pasture, so there are hedgerows or forests around each field. This provides a windbreak during cold weather and some shade during hot summer days.

Brooke and Tom invested in a high-quality, low-impedance energizer and a multistrand high-tensile

Figure 1.3. A single strand of polywire on a reel with some portable posts makes it possible to move the beef herd at Spring Pastures Farm once or more each day. This type of portable fencing also makes it easy to change paddock size as needed.

Figure 1.4. Tom and Brooke move the herd to a new paddock that has transformed to high-quality forage after less than three years of careful management.

electric fence around the perimeter of all the pastures. To get the energizer to work well, they had to put in 11 ground rods. This large ground system is typical for this type of energizer; indeed, an undersized ground is a common beginner mistake that Tom and Brooke wanted to avoid. The energizer's remote shut-off allows them to turn it off from wherever they are on the farm. This makes it easier to move or repair fence; there's no need to walk back to the energizer to turn off the fence.

For interior paddock subdivisions, they use a single strand of polywire with portable posts. (For details on types of electric fencing and how to design a system, see chapter 14.)

Buried water lines run to hydrants in each pasture, supplying portable tubs to provide drinking water to the herd. Thus the animals can remain enclosed in each area instead of having to walk from the paddock to a central water source.

CHAPTER 2

Types of Grazing Systems

With all the benefits of pasture-based farming, it is no surprise that there is considerable interest in grass farming. In the past, it took quite a bit of effort and research to find useful and accurate information on how to set up a well-managed grazing system. Now information on grass farming is plentiful. In fact, it is so abundant that it can be confusing to someone new to the topic. Choosing the "best" grazing system to set up, or the "best" way to make improvements to an existing system, becomes easier with an understanding of the range of grazing systems available. It also helps to learn the basic guidelines that determine whether a grazing system will be beneficial and successful. Methods of pasture management include:

- Large continuously grazed pasture systems.
- Simple rotational grazing systems.
- Intensively planned and managed grazing systems.

Each type of system requires different amounts or types of fencing, livestock drinking water systems, labor, and management. Each of these will also create big differences in the quantity and quality of feed and the health of the pasture ecosystem.

Continuous grazing, also called extensive grazing, is the simplest and most common grazing system. Rotational systems can be an improvement on the continuous system if they are carefully managed. Better grazing systems include Management Intensive Grazing (MiG) and Holistic Planned Grazing (HPG). These more intensively managed and planned systems create healthier grazing ecosystems because they are based on a thorough understanding of the key guidelines of pasture management including variable recovery periods. Table 2.1 allows a quick comparison of some of the common types of grazing systems.

Continuous Grazing

Continuous grazing systems are those where livestock are allowed to graze the same pasture for most or all of the grazing season. Such a system requires the least day-to-day management of all grazing systems, but it usually provides the animals with a less consistent supply of high-quality feed, and may result in lower growth rates and milk yields than are possible using better management systems.

When livestock have continuous access to the plants in a pasture, overgrazing damage of plants occurs. Because of this damage, continuously grazed pastures are more likely to have low plant vigor, low plant density, poor soil health, and a higher risk of erosion. For this reason, continuous systems require more acreage of pasture per animal, more clipping of pastures, and possibly occasional pasture reseeding and renovation over time.

To understand the problems caused by continuous grazing, it's important to consider the situation from the plants' perspective. When an animal grazes a plant, it removes some of that plant's foliage. Pasture plants need time to rest after being grazed in order to photosynthesize, regrow their leaves, and replenish their energy reserves. When animals are continuously grazing in the same pasture, or animals return to a pasture before it is fully regrown, plants lose out on that essential recovery time, and this causes damage, such as excessive loss of leaves, from which they may not be able to recover. Continuous grazing also gives cattle time to graze the plants down very short, which can cause damage as well, particularly to the plants' growing points. The resulting weak plants may stop growing or die. These weakened, overgrazed plants will not compete well with weed species or hold soil well. The result can be bare soil and erosion.

Figure 2.1. One result of damage from continuous grazing with a high stocking rate can be pastures like this one. The author's hand provides a height gauge, showing that the pasture is full of very short, nonproductive plants even though it has been regrowing for over a month.

Figure 2.2. There appears to be plenty for the cows to eat in this pasture, which has been continuously grazed for several years at a low stocking rate. However, a close examination would show areas of very short overgrazed plants, other patches of overmature, unpalatable plants, and an increasing number of woody and weedy species.

Under this type of poor grazing management, where plants are not allowed to fully recover between grazings, there are more likely to be patches of bare soil, less desirable grass and clover plants, and increasing weeds. Patches of clover, mosses, or grass will appear that never grow very tall, while in other areas clover and other desirable pasture plants may be completely absent. Where livestock reject weeds or unpalatable plants, brush or taller overmature pasture plants may start to appear. There may also be a buildup of dead plant material or thatch on the soil surface, along with manure that has not decomposed quickly. These are all symptoms of damage from poorly managed grazing.

Simple or Fixed Rotational Systems

In a simple rotational system, the livestock are moved periodically among a few pastures, often on a fixed rotational schedule where the herd or flock is moved every few days or once a week. The benefit of the simple rotation is that it affords pasture plants a few opportunities during the grazing season when they can regrow.

Figure 2.3. In pastures grazed for several weeks then rested for a week or two throughout the season, there will be areas where more productive pasture plants have been lost, replaced by shorter, less productive species. As animal trails form, weed species will take over other areas.

Table 2.1. Comparing Grazing Systems

System Characteristic	Continuous Grazing Systems	Rotational Grazing Systems	Management Intensive Grazing and Holistic Planned Grazing
Period of occupation and frequency of moves to new pasture	Livestock remain in the same paddock for the whole grazing season.	Livestock rotate around several pastures, often on a set rotation. Recovery periods are kept the same length, even when plant growth slows.	Livestock are moved to a new paddock only when it has fully regrown. Recovery periods are variable based on plant regrowth time requirements.
Forage supply	Livestock graze selectively, making it difficult to balance the ration. Pastures will generally provide enough feed in spring, but later in the summer pasture will be too short or too mature to provide enough dairy-quality feed.	Livestock may have adequate dry matter intake (DMI) and pasture quality in the spring, but as plant growth rates slow the pastures will be too short or plants will overmature and provide lower-quality feed.	Livestock will have adequate DMI and pasture quality throughout the grazing season.
Forage quantity	Pasture quantity usually declines as the season progresses. Productivity of the pasture also declines from season to season as plants become more damaged by overgrazing.	As livestock rotate back into pastures that are not fully regrown, the quantity and quality of feed will decline. Each year the productivity will be lower.	Livestock only rotate back into pastures that are fully recovered, so pasture quality and quantity remain good. More acres are added into the grazing rotation as growth rates slow. Over the years, this management system will increase pasture productivity.
Forage quality	Pasture quality will decrease each year due to overgrazing damage, increased weeds, and rejected forage. Clipping and eventually renovation and reseeding may be needed.	Pasture quality will gradually decrease due to overgrazing damage, weeds, and rejected forage. Clipping can help prevent weeds from spreading, but eventually renovation and reseeding may be needed.	Pasture quality will improve over time. The more intensive the management, the faster the pasture will improve.

However, these systems are not designed to allow the plants sufficient time to regrow after each grazing, and plants may also be grazed down too short.

Shifting from a continuous system to a simple rotational system can make it possible to improve pasture quality and quantity to a limited extent. However, pasture productivity will still be lower in a rotational system than in a more intensively planned and managed system.

Intensively Planned and Managed Systems

By *intensive* we are not talking about how short the pasture is grazed down. *Intensive* refers to the management itself. Successful systems that are more management- and planning-intensive provide livestock with new areas of high-quality pasture frequently, followed by a long rest period to give those areas of pasture time to regrow before the next grazing. We could also call this controlled or carefully managed grazing.

These more carefully planned or management-intensive grazing systems are designed to take care of pasture plants and soils. This results in more high-quality forage, many environmental benefits, and better livestock performance. A fundamental difference between a simple rotational system and a higher-quality, intensively planned and managed system is that the latter is managed by paying close attention to how fast the plants in the pastures are growing. As plant growth rates slow, the farmer increases the length of the recovery period after each grazing period to make sure plants are always fully recovered before the next grazing. This key principle of *variable recovery periods* is essential to create the highest-quality pastures.

Over the years, many terms have been used to describe the types of grazing systems that are based on high-quality management. At workshops and field days, people sometimes ask me, "Does the name of the grazing system matter?" My response is that the *names* given to a grazing system are far less important than the *differences* in how the system is planned and managed for the care of plants, soils, and livestock.

André Voisin used the term *Rational Grazing* in his writing in the 1950s. Some farmers call this type of grazing system Voisin Grazing. Voisin, who was both a scientist and farmer, described some basic guidelines necessary for good grazing management. These included sufficient regrowth periods following each grazing, short periods of occupation when grazing each paddock, and allocating the best pasture to the livestock with the highest nutritional needs. His writing on careful management of soils and plants described in detail why it is important to manage how much energy is stored in the lower parts of the plant.[1]

> A pasture plant must be capable of growing again after it has been cut either by the tooth of the animal or by the blade of the mower. When this plant is cut it retains very little, and sometimes indeed hardly any of the green aerial part capable, by photosynthesis, of creating the elements necessary . . . for the initial regrowth of the plant. It is then indispensable that the plant, at the moment when it is cut, should have, in its roots or at the foot of its stalks, sufficient reserves to allow the formation of a certain green portion which, by photosynthesis, will then permit the normal growth of the plant.[2]

More recently, grazing consultant and author Jim Gerrish started using the term *Management Intensive Grazing* (MiG).[3] I like MiG as the emphasis is on the importance of the *management*, which is why the "M" is generally capitalized while the "i" is lowercase. Gerrish sometimes quotes an even earlier author on the subject of good grazing management, James Anderson, who published essays on the topic in the late 1700s!

Figure 2.4. This flock at Vermont Shepherd Farm enjoys high-quality pastures created by using short periods of occupation and long recovery periods between each grazing. Dairy animals require particularly high-quality pasture (dairy quality!), which on this farm includes white clover, dandelion, Kentucky bluegrass, and many other perennial species of legumes and grasses.

Allan Savory's Holistic Planned Grazing (HPG) is part of a comprehensive Holistic Management planning system, which includes financial planning, biological monitoring, and establishing goals.[4]

Mob grazing is another common term; it refers to use of high stocking density with long recovery periods. However, farmers using MiG or HPG may also use high stock densities and long recovery periods.

Another term used in many Natural Resources Conservation Service (NRCS) publications is *prescribed grazing*. Both MiG and HPG, when done correctly, will meet the NRCS's prescribed grazing definition. There is additional historic information on grazing methods and terminology at the beginning of chapter 5.

The names applied to each of these systems are interesting, but what's important is understanding the key principles all successful grazing systems share and how to apply those principles on your own farm. Whatever the name, a system that is well designed, planned, and

KEY PRINCIPLES OF GOOD GRAZING MANAGEMENT

Well-designed and well-managed grazing systems will be most likely to both meet livestock needs and improve the productivity of the pasture plants and ecosystem if they are based on these two important guidelines:

- Graze livestock in each area for a relatively short time (short period of occupation) to prevent regrazing.
- Allow plants as much time as they need to fully regrow and recover after each grazing (variable recovery period).

managed pays close attention to the needs of the plants, livestock, and soils. This approach favors the better pasture plant species, reduces weed problems, improves soil health, and increases the quantity of available forage produced while improving the nutritional quality of the feed.

What Does Quality Pasture Look Like?

High-quality pasture should contain a diverse mix of many perennial plant species that form a dense stand of plants with no bare soil, as in figure 2.5. Manure from the most recent grazing should show uniform distribution. There may be patches of plants that were not closely grazed during the last grazing, since grazing livestock do not like to eat the grass right next to their manure. This

Figure 2.5. Cows at Kriemhild Dairy in New York are able to fill their rumens quickly on this high-quality pasture before lying down to chew their cuds.

high-quality pasture will produce feed with the highest nutritional value and vitality so that animals are healthy and the meat, milk, and manure produced are of the greatest benefit to the farm.

This book focuses on the more intensively planned and managed systems of grazing, because such systems generate higher yields and better-quality pasture. It is possible, however, to use information on the fundamental principles of good grazing management to set up a less intensive system that still provides enough pasture to meet farm goals. In this case, additional acres per animal will be needed, because grazing efficiency and pasture productivity will be lower. It will still be necessary to add land to the grazing rotation as plant growth rates slow in summer.

The design of the grazing system should be determined by the farm's overall goal and the production objective for the livestock. What works for one farm may not work on another. For example, a farmer with a full-time off-farm job may not be available to move the cattle herd to a new paddock twice a day. However, a dairy farmer whose goal is to maximize dry matter intake (DMI) from pasture may want to move the herd to fresh pasture three or more times a day. Both farmers will benefit from having a well-designed and -planned grazing system based on a solid understanding of the needs of the pasture plants and grazing livestock.

PART TWO

Grazing from the Plant's Perspective

For better or worse, the productivity and types of plants growing in a pasture are there as a result of past management decisions by the farmer. By gaining a deeper understanding of how pasture plants prefer to grow and how to build healthy soils — as well as a deeper appreciation of the incredible responsiveness of the pasture ecosystem — it is possible to use livestock grazing to create a highly diverse, productive, high-quality pasture. There are many different types of grasses, legumes, and other plants that grow in pastures. By learning more about how each of these species responds differently to different types of grazing management, weather, and soils, we can further improve pasture health and vigor. This can allow us to capture as much sunlight as possible in our pastures to improve our environment, pasture productivity, animal welfare, and human health.

CHAPTER 3

Grazing-Adapted Plants

Pasture plants are part of a complex ecosystem. They are affected not only by grazing management, but also by natural disturbances, wildlife, insects, disease, and weather fluctuations. It can be difficult to determine whether changes observed in a pasture are caused by the farmer's grazing management decisions or by other factors such as weather or soil conditions.

Some plants are well adapted to grazing, and will thrive under good pasture management. Other plants prefer not to be defoliated by livestock, making them less adapted to grazing. Learning more about how different pasture plants grow and how they respond to different types of grazing can help us better design and manage grazing systems.

The green plants in our pastures have an incredible talent. They turn sunlight into substance. Through photosynthesis, plants are able to take energy from sunlight and carbon dioxide from the air to make carbohydrate sugars, which are then stored in the roots and lower parts of the plant stem. They also release oxygen into the air and store carbon in the soil (carbon sequestration).

As plants grow, minerals taken up from the soil by roots are combined with the photosynthetic carbohydrate to make fiber, protein, and fats. This process is what supports life on earth! The process of pasture plants converting sunlight into an energy-rich source of food is what makes it possible to be a grass farmer.

The more healthy green leaves that pasture plants have, the more photosynthesis can occur. Maximizing the amount of photosynthesis taking place in pastures is one of the most important jobs of a grass farmer. However, it is also important to manage how much, and how well, those photosynthetic carbohydrates are stored in the plants.

Stored carbohydrates are what allow perennial pasture plants to live through the winter, break dormancy

Figure 3.1. The energy in the sunlight that lands on the green leaves of plants is captured by the photosynthetic process. In this process the plants take carbon dioxide from the air, release oxygen into the air, and store energy in the form of carbohydrates in the plant. Illustration by Anna Powell.

Figure 3.2. The farmer is moving the beef herd out of this paddock before they consume too much leaf area. With so much residual growth left in place, the plants should regrow more quickly.

Figure 3.3. In this pasture, by contrast, most areas were grazed down much shorter, and the plants have a lot less leaf area. These plants will be more stressed, and will regrow roots and leaves more slowly. They'll need more time to regrow before the pasture can be grazed again.

in the spring, and begin growing again. This stored energy is also needed so they can regrow after they are grazed. These carbohydrate energy reserves are essential for plant survival and health, because they are the only energy source the plant has for regrowth after severe defoliation or in the spring when green leaves are not present.

Plants won't start storing new energy until they have generated new leaves and had enough time to photosynthesize. This regeneration process can take as little as two weeks or as long as a few months. Allowing animals to graze plants before the plants have refilled their energy reserve damages the plants. The more often plants are grazed before they are fully regrown, the more severe the damage will be. This is why making sure the regrowth period is long enough following each grazing is so important.

Plants use two sources of energy to regrow: existing healthy green leaves, which are able to make carbohydrates by photosynthesizing; and carbohydrates stored in the lower parts of the plant. If enough residual plant material with green photosynthetically active leaves is left in the pasture after grazing, plant photosynthesis will be able to provide most of the needed energy for regrowth. However, if the pasture was grazed short, leaving little residual, then plants must move stored carbohydrates

from their lower parts up to grow new leaves. The worst damage to plants is done when they are first grazed very short by the livestock, and then not given enough time to fully regrow before being grazed again. This double damage will rapidly deplete plants' energy reserves, resulting in lower plant productivity and plant death.

Grazing pasture plants before they are fully regrown, or grazing them down too short, leads to a significantly less efficient and productive pasture system, because what happens to the plant tops also affects root growth. Plants will have the most vigorous root growth if less than half their leaf area is removed at each grazing. A plant that has more than half of its leaf area grazed off will slow or stop the growth of roots and will take longer to fully regrow both its leaves and roots.

The Importance of Regrowth Periods

The length of time that plants need to regrow varies depending on the type of plants, season, temperature, and rainfall or irrigation pattern. For example, in some locations during rapid spring growth, pastures will regrow in just three weeks. However, during midsummer heat and dry conditions, those same pastures may need two months or even more before they can be safely grazed again. Thus, when spring pasture growth slows

AVOID THIS COMMON MISTAKE

One of the most common mistakes in a simple rotational system is not reducing the speed of the rotation as plant growth rates slow. If livestock return to a pasture before it is fully regrown, the plants won't have had time to recover and there will be overgrazing damage. This is also why continuously grazing animals in the same pasture causes damage.

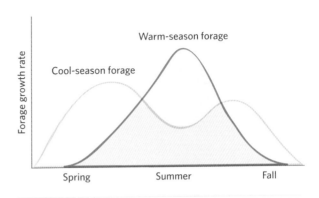

Figure 3.4. The two curves in this graph show that cool-season and warm-season plants have their season of rapid growth at different times of the year. Farms that are able to grow both types can create a longer grazing season with a more stable forage availability.

during hotter, drier times of the grazing season, the pastures must be given longer to regrow.

This midsummer slowdown can create what is often referred to as a forage gap or midsummer slump — a time when there isn't enough pasture growth to keep up with livestock nutritional needs. Several strategies can be used to bridge this forage gap and increase the length of the recovery period. These include adding more acreage into the rotation, decreasing herd size, grazing forages that grow more rapidly in midsummer heat, and restricting access to pasture while increasing supplemental feed in the barn.

In northern regions where summers tend to be cooler, cool-season grasses are among the most common pasture plants. These perennial plants are more productive in cooler weather with moist conditions, so they will be most productive in the cool spring and fall. They produce a large volume of forage in the spring, but particularly in hot, dry weather there may be little or no growth — hence the midsummer slump. Another category of plants, commonly found farther south in warmer climates, are warm-season grasses. These plants are more productive during the hot, dry midsummer weather.

Knowing what types of plants are already growing on the farm, as well as understanding under what conditions they will grow most vigorously and how to manage them, can help manage the challenges of the midsummer forage gap. For example, some farms grow separate fields of warm-season perennials or annuals for grazing in midsummer. Other farms have warm-season grasses mixed in with the cool-season plants. Warm- and

Figure 3.5. This overgrazed pasture has many patches of plants that have adapted to repeated grazing pressure by staying very short. The taller plants are those that the cows always refuse to eat. Weeds such as buttercup can take hold because animals don't find them palatable.

cool-season grasses are discussed in more detail later in this chapter, and in chapter 6.

The most common way to compensate for a forage gap at times of the year when pasture growth rates slow is to add more pastureland to increase the total number of acres in the grazing rotation later in the season. This is usually done by taking an early cut from a hay field

Figure 3.6. Unpalatable species like these thistles don't get grazed, so they thrive in an overgrazed pasture.

Figure 3.7. This pasture includes several species of cool-season grasses and clover. Due to both the density of plants and the diversity of leaf types, most of the sunlight that lands on this pasture is usable.

and then adding that to the grazing rotation once it has regrown. Chapter 5 discusses variable recovery periods in more detail.

If the number of grazing acres remains constant as plant growth rates slow and none of the other strategies are used, the plants will not get enough time to regrow after each grazing. Due to the plants not being fully regrown and therefore shorter, there will be less forage available in the pasture, and dry matter intake by animals will drop. So both animals and pastures will suffer due to this grazing management mistake!

Weakened overgrazed plants will not compete well with weed species. They will not hold soil well, which can result in bare soil and erosion. Some grasses and clovers will survive by staying very short, never growing tall enough for livestock to graze easily, while livestock

will reject other areas that will soon grow up into less palatable plants, weeds, brush, or small trees.

Maximizing Sunlight Conversion

The amount of sunlight energy that pasture plants are able to convert into plant growth depends on how much light the leaves can intercept, and how efficient photosynthesis is. A dense pasture canopy made up of many different plant species that bear healthy green leaves will capture more light and be more efficient at energy collection and storage.

The light that reaches an individual plant leaf varies in quality and quantity with latitude, elevation, season, and atmospheric conditions. Taller surrounding plants are also a factor because they filter the light. The

Figure 3.8. This pasture has low diversity and low plant density. It does not intercept as much sunlight as a dense, diverse stand of pasture plants can.

percentage of light above the plant canopy that reaches within the canopy decreases as the pasture grows taller and denser. Not only does less light reach the lower areas under the pasture canopy, but the spectral quality of the light changes.[1] This means that "shade" isn't just a reduction in the total amount of light; it's also a change in the quality of light.

A good way to understand this concept is to take a walk in a forest and watch how the light passes through the canopy of tree branches and leaves. You'll see a few patches where sunlight makes it through to the ground, but most of the forest floor will be in varying degrees of shade. This is similar to the quality of light that reaches the ground surface in a highly dense pasture.

As light passes through a plant canopy, blue and red wavelengths are absorbed but far-red wavelengths are transmitted, so far-red light makes it farther toward the ground. The reduction in the ratio of red to far-red lights leads to changes in plant growth in underlying shaded areas of the pasture. Plants such as white clover will produce fewer roots and branches along their stolons (stems that grow horizontally on the soil surface). In addition, clover plants will grow longer leaf petioles and stolons will elongate, allowing them to reach areas in the pasture where sunlight is more available. As shading effects become more severe, and less photosynthetically active light reaches the plants in the lower area of the canopy, plant growth will slow due to less carbohydrate availability.

Leaf area index (LAI) is the surface area of leaf blades per unit of ground area, and this measurement tells us how much light the pasture will intercept. More diversity, particularly a mix of legumes, grasses, and forbs, will increase the LAI and amount of light the pasture can intercept. Increased plant density also increases LAI. Higher LAI allows a pasture to capture and store more energy so plants grow faster and produce more forage.

Grazing management strategies that allow pastures to become more diverse and encourage higher plant density not only increase LAI and photosynthesis, but also allow the grazing livestock to consume more forage per bite.

However, not all pasture plant parts photosynthesize with the same efficiency. Several factors influence the efficiency of photosynthesis. One of the most important factors is the age of the leaves, because younger leaves have higher efficiency.[2] Thus, managing to maintain more vegetative leaf growth can also increase photosynthesis. A pasture that has passed into the reproductive growth phase will have more stem and seeds with fewer vegetative leaves, and therefore be less efficient at photosynthesis.

As noted on page 64, short-duration grazing with a high stock density, followed by enough time (but not too much time) for plants to regrow, is an effective grazing management strategy to increase density, diversity, and vigor of pasture growth and maximize sunlight capture. This is covered in more detail in chapter 5, including discussion of how to accomplish this with different types of plants (cool-season grasses, warm-season grasses, legumes, or forbs) and at different times of year.

GRAZING IMPACT ON ROOTS

It is easy to focus on just the aboveground parts of the pasture, since that's what we look at most of the time. However, these leaves and stems aren't the only plant parts that grazing affects. Like leaves, the roots of pasture grasses also grow and then die, but roots tend to live much longer than leaves. The diversity and types of plants, as well as the method of grazing management, has a large impact on root health and vigor.

Roots do a lot of important work for the plant, and for soil health. Roots anchor plants in the soil, and they take in water and minerals for plant growth. If roots are reduced due to poor grazing management, plants will be less able to take in water and minerals during droughts, and plants will be less likely to thrive during poor growing conditions. Overgrazed plants with weakened roots are also less able to hold the plants and soil in place to prevent erosion.

In addition to capturing more sunlight, maintaining a diverse mix of pasture plants provides livestock with many options to choose from when grazing. Plant diversity also assures that some growth will occur even when weather conditions are extreme. And because different plants grow differently throughout the growing season, diversity can also increase the length of the productive growing season.

Managing pastures to encourage diversity of desirable species requires a thorough understanding of how different pasture plants respond to grazing. The better planned and managed your grazing system is, the more rapidly the pasture will develop higher plant density, improved plant diversity, and more vigorous growth throughout the growing season, which in turn leads to improved ecosystem health, pasture productivity, and forage quality.

Types of Pasture Plants

Well-managed perennial pastures provide livestock with a mixture of many types of plants, providing a variety of nutrients and medicinal compounds. In *Reproduction and Animal Health*, Gearld Fry stated: "It is no accident that good pastures have perhaps 40 to 50 species . . . Nor is it an accident that fully half the weeds in the USDA index are also listed in manuals on medicinal plants."[3]

John Hendrickson and Bret Olson wrote in *Understanding Plant Response to Grazing* that "landscapes are collages of complex plant communities and site conditions."[4] Many different plants make up this complex pasture community, and they don't all follow the same patterns of growth. It's helpful to understand how each distinct plant type is most likely to thrive in the pasture.

Pastures are made up of grasses, legumes, and forbs (and some weeds), and they all have somewhat different growth habits. Some are shorter, some grow taller; some branch and spread horizontally while others grow in upright bunches. Tolerance to dry or wet soils and to hot or cold temperatures also varies. Some plants are well adapted to grazing because they can recover quickly after grazing or defoliation, while other plants are able to tolerate only light and infrequent grazing. Certain grazing-adapted plants will tolerate being grazed quite short, while others require less severe grazing in order to persist in pasture. However, even a plant that is well adapted to grazing will be damaged if it is grazed too severely or too often.

Response to grazing doesn't happen in isolation, because plants compete with neighboring plants for water, sunlight, and nutrients. For example, if a plant has been grazed more severely than the plant next to it, it will be at a disadvantage in terms of regrowth. How quickly a plant recovers from grazing depends largely on how quickly it can reestablish leaves and start photosynthesizing. Most plants won't be killed by a single grazing, but some can tolerate many more grazings and more severe grazings than others.

Healthy roots help plants tolerate grazing, and under good growing conditions a plant that is well adapted to grazing can restart root growth within a few days of being grazed. However, if a plant has been grazed severely, it won't restart root growth as quickly, and thus can't take up water and nutrients as well as its neighbor. The location of the plant's growing point — the site where new leaves or stems develop on a plant — can determine how well the plant can tolerate grazing. For example, many of the growing points of grasses are near the base of the plant, while shrubs bear their growing points at the tip of each shoot. Thus, the growing points of grasses are less likely to be completely bitten off by the animals, simply because they are harder for the animals to reach. Because the growing point remains intact, grass is better able to regrow quickly. However, when an animal such as a goat grazes a shrub or other species with elevated growing points which are easier to get to, it's likely to nibble off many growing points, making regrowth much slower.

Even among grass species there are differences. Sod-forming perennial grasses such as Kentucky bluegrass have growing points and crowns that are extra low to the ground, so they can tolerate being grazed short. This is why bluegrass is commonly found in horse or sheep pastures where the animals have regularly grazed the plants very short. Plants with crowns and growing points that are low and hidden from the livestock survive better. The crown, which is the basal portion of the grass plant, is essential to the growth of perennial grasses. This area is where overwintering tissues, energy storage, and growing points are found, which in the spring or after a period of dormancy produce new growth and shoots.

In contrast to Kentucky bluegrass, the crown or growing points on grasses such as timothy or orchard grass are positioned higher above ground level. This makes these grasses less tolerant of very short grazing. If livestock eat the growing points and crowns too often or severely, those plants will die and be replaced with ones that can tolerate severe grazing.

Growing points are found in several different locations in pasture plants. Some are found on the tips of roots and stolons (apical growing points). Growing points called basal buds are found in the crown of the plant, and allow tillers or new stems to emerge. Certain grass species also have aerial buds, which are found at some of the lower nodes of the stem. Many grasses also have growing points called intercalary meristems at the base or collar of the leaf blade. These are the source of continued leaf growth from the base of each leaf. See figures 3.9 and 3.15 for more information.

Palatability also varies among different types of plants. Some species are less likely to be grazed as severely or as frequently as others due to livestock feeding choices. Some plants have low palatability due to high fiber or physical barriers such as spines and thorns, which make it difficult for the animal to graze it. Plants also contain secondary compounds that may reduce the forage quality or make animals less likely to graze them. These compounds include tannins, alkaloids, oxalates, glycosides, and terpenes. These can deter grazing by lowering plant digestibility, and some may be toxic.[5] (This is discussed in more detail in chapter 10.)

Livestock are selective grazers, which means they choose to graze plants and the parts of plants that are more desirable and palatable. Livestock, similar to humans presented with a buffet featuring a variety of foods, are likely to eat their favorites and leave the rest on the table.

When livestock are given a choice of many plants in a large pasture, this can lead to uneven grazing. Certain plants will be stressed or overgrazed while others are left untouched and can continue to grow. Selective grazing can be a significant issue in pastures that are grazed with lower stocking densities. When livestock are given a larger area to graze, during a period of days or weeks, they will wander around picking and choosing what they want to eat. Increasing the stocking density of animals in the pasture and frequently moving them from one paddock to another within the pasture reduces this selective grazing by limiting the amount of time the livestock stay in any single area and restricting the relative availability of forages that may be more desirable. Figure 2.2 shows an example of selective grazing with low stocking density in a continuously grazed pasture. Weeds and unpalatable plants are abundant and healthy thanks to livestock choosing not to eat them. Figure 3.5, by comparison, shows the effect of selective grazing with a higher stock density. Most of the pasture is very short, but there are rejected patches of the most unpalatable weeds.

How Plants Like to Grow

Perennial and annual grasses, legumes, and forbs all have different growth habits, and growth habits also vary even within the grasses group. Understanding the differences in how plants like to grow helps us design grazing systems that encourage plants to thrive, build healthy soils, and produce quality feed for livestock. Here, I'll present an overview of plant types and growth habits. For more detail on individual plant species, refer to chapter 6.

Grasses

Grasses include annuals, which live for just one season, and perennials, which are plants that live for more than two years. Let's start with the perennials, and discuss the annuals later in this section.

Perennial pasture grasses include both warm- and cool-season grasses. Cool-season perennials like to grow during cool, moist conditions, so they will be most productive in the cool times of spring and fall. These plants will produce most of their forage dry matter at the beginning of the grazing season. Some cool-season varieties, particularly in locations with hot dry summers, may go dormant for a period of time during midsummer. Warm-season grasses, by contrast, grow best and are more productive during hot, dry midsummer weather. Warm-season grasses can produce very high yields of forage for both grazing and harvested feed in the heat of summer.

Perennial grasses include species, such as Kentucky bluegrass, that are very well adapted to grazing. However, other perennial grasses such as smooth bromegrass or timothy are more susceptible to being grazed down too short. As discussed earlier, this is due to differences in how the plants grow. Learning about plant anatomy and physiology can help in understanding why some grasses do better than others in the pasture, and why some respond very differently to different grazing techniques.

When you look closely at the anatomy of a grass plant, you will notice that each plant is a collection of tillers — individual plant shoots — that grow from the base or crown. (See figure 3.9 for a visual example of grass plant anatomy.) Tillers can develop from seeds, they can shoot out from the base of existing grass plants, or they can arise from stolons or rhizomes.

Each tiller consists of at least one leaf as well as the growing point. The growing point for each tiller is at or near the ground level. As the tiller begins to grow, additional leaves will emerge. In the spring after dormancy, new growth will be initiated from the tiller's ground-level growing point. Later, as the tiller stem elongates, it will be easier to differentiate the nodes and internodes, and to see that each leaf attaches to the stem at a node. As the stem elongates, the internodes (stem areas between the nodes) become longer and some new growing points will be found higher up on the plant stem and leaf collars.

On each grass tiller, the older leaves are at the bottom; newer leaves are above them at the top. As a newly sprouted tiller grows, it will develop its own root system and can eventually sprout more tillers. Thus, each tiller has the ability to become another self-supporting grass

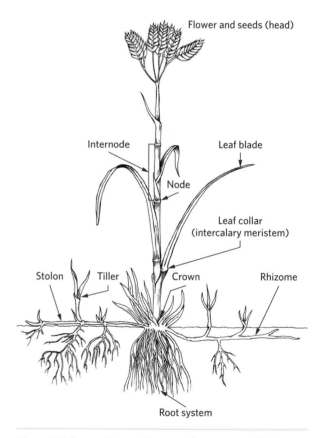

Figure 3.9. Depending on the type of grass, new tillers may develop from seeds, roots, or stolons, or from the base (crown) of a grass plant. Each grass tiller can eventually become a new plant. Illustration by Anna Powell.

Figure 3.10. This orchard grass plant has new tillers emerging through dead leaves left over from the previous year's growth.

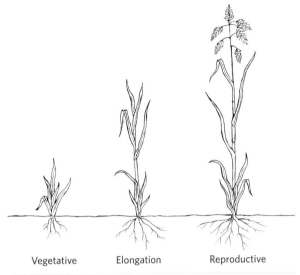

Vegetative　　Elongation　　Reproductive

Figure 3.11. During the early vegetative stage, the plant grows only leaves while the growing point stays close to the ground. When the stem begins to elongate, it is possible to more clearly see that each leaf is growing out of a node as the internodes elongate. Leaf blades can continue to elongate from their growing points in the collar or intercalary meristem. Once the reproductive stage is reached, a flower and then a seed head will form at the upper end of the stem. Illustration by Anna Powell.

plant if it is separated from the rest of the mother plant. As long as the new tillers remain connected to the rest of the grass plant, they can share stored energy reserves while they regrow after being grazed. This is one of the characteristics that helps grasses adapt to grazing.

The basic physiological stages of grass development include the vegetative stage, the elongation stage, and the reproductive stage. If the grass plant is growing from a new seed, there will also be an earlier stage: germination. The germination stage begins when temperature and moisture conditions in the soil make it possible for the seed to germinate. This is the stage in which the first shoot emerges from the seed.

During the early vegetative stage, the growing point remains compact near the soil line, where the crown of the grass plant is. This vegetative stage is when leaf growth and development occur. Later, the internodes on the stem will begin to elongate (this is also called jointing). This is the stage when the stems grow taller and become more visible in the pasture. This also elevates

some of the growing points from near the soil level to higher up in the pasture canopy. This can be seen in figure 3.11.

In addition to the growing points found in the crown where new tillers or stems form in these elongating grasses, there are also growing points at the collar or base of each leaf. The tips of the leaves are *not* where the growing points are found! When the leaf tip is removed, the tip doesn't regrow. Instead, cells at the base of the leaf (at the collar or intercalary meristem) elongate to increase the length of the leaf blade.

Growing points may also be found in other locations in grasses, and may also be called buds. Buds at the crown of the plant produce new tillers. Buds may also be found on rhizomes and stolons. On some species, buds may also occur at nodes on the lower parts of the elongating grass stem.

After the growing point on a tiller changes to reproductive mode and starts to form the flower head, it will not produce more leaves. Instead, the stem will elongate

Figure 3.12. In early spring these grasses are in the vegetative growth stage, so the growing point remains at the base of the plant where it emerges from the soil.

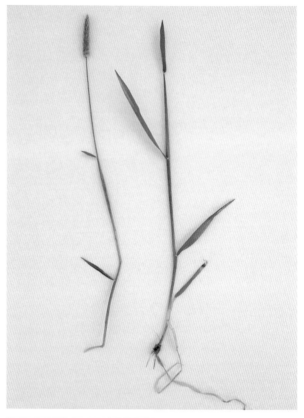

Figure 3.13. The grass plant on the right is elongating. On the grass plant to the left, the flower has already emerged.

to raise the seed head up. The flower will emerge and seeds begin to form. The final part of the reproductive phase occurs when the seeds ripen and eventually are fully developed.[6]

When a grass plant is just beginning the early vegetative stage, it doesn't have much leaf area, and the rate of growth will be slow. But as it starts to generate more green leaves, it can capture sunlight and grow faster. The fastest growth rate occurs in later vegetative stages. Once the stem has elongated and the plant shifts into the reproductive stage, the dry weight of the plant doesn't increase; much of it is just redistributed between the different parts of the plant.

To judge when the plants in a pasture are ready to graze, it is helpful to be able to distinguish whether they are in the vegetative stage, making lots of highly digestible leaf, or if they are instead making a lot of less

digestible stems as they get ready to flower and make seeds. This allows the farmer to make good decisions that care for the plants, which need enough time to grow, as well as for the cattle, which need high-quality digestible forage.

There are some additional categories of grasses that have different growing habits. These include sod-forming grasses, bunchgrasses, and grasses that elongate their stems even when they are not in the reproductive stage. Understanding these different growth habits makes it easier to observe what is happening in the pasture and make good management decisions.

Bunchgrasses include orchard grass, big bluestem, and tall fescue, while sod-forming grasses include Kentucky bluegrass, reed canary grass, and quack grass. When these species are present in a pasture together, each may be ready to graze at a very different height.

Figure 3.14. In this Maryland pasture, bunchgrasses are surrounded by lower-growing sod-forming grasses and clover. Cows are able to graze a variety of taller bunches and shorter pasture plants.

Some grasses, such as Kentucky bluegrass, keep their growing points at or below the ground level throughout the season. Others, such as timothy, reed canary grass, smooth bromegrass, and switchgrass, elongate their stems and elevate their growing points even when not yet in their reproductive stage. Because these plants produce more stem, they tend to produce less digestible forage. In addition, because they elevate some of their growing points, grazing is likely to remove growing points from these plants. Thus, they also require a longer period between defoliations so that they can regrow new leaves, photosynthesize, and replenish their stored energy reserves. These types of "jointing" or "elevating" grasses are thus less well adapted to grazing. They can still thrive in a well-managed pasture but require careful management so they are not grazed too short or too often.

Cool-season grasses with higher growing points (like timothy) will do best if animals don't graze the bottom 3 to 4 inches. Some warm-season grasses need even more of the lower grass stem left behind after each grazing. By learning what types of plants are growing in the pasture, and how they each prefer to be grazed, the farmer can manage to improve those plants' health and productivity.

Many of the grasses produce new tillers in the late summer and fall. These will be productive new grass plants and tillers in the following grazing season, so careful grazing of them in the fall will increase pasture productivity the next year. Leaving an adequate post-grazing plant residual, and giving the grasses enough recovery time to grow multiple leaves on these new tillers before the first killing frost, will allow them to build up a larger energy reserve with which to begin growth next season. Conversely, overgrazing those plants in

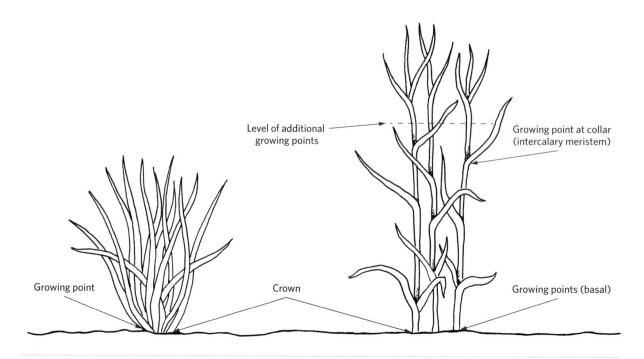

Level of additional growing points

Growing point at collar (intercalary meristem)

Growing point

Crown

Growing points (basal)

Figure 3.15. The grass on the left has its growing points at the soil surface. Some growing points on the grass on the right are now found higher up the elongating stems. Illustration by Anna Powell.

the fall will reduce how productive they are in the next growing season. Each grass tiller has significantly more nodes for roots than for leaves. An individual tiller may have three to five leaves, but it may have 10 root nodes. Despite this, the amount of energy from photosynthesis channeled to the roots is relatively small compared with that devoted to the rest of the plant. Younger roots receive more energy than older ones. Less energy is directed to the roots during seed head development, after a summer drought, and following grazing.[7] This is yet another reason that regrowth periods must be long after each grazing.

Cool-Season or Warm-Season?

As mentioned earlier in this chapter, cool-season species are more productive in cooler weather with moist conditions. Some common cool-season perennial grasses suitable for grazing include orchard grass, Kentucky bluegrass, and perennial ryegrass. Warm-season grasses are more efficient at gathering carbon dioxide while using less water, which is why they can be more productive during hot, dry weather. Some warm-season perennial

pasture grasses include switchgrass, big bluestem, and annual species such as sudan grass, corn, and millet.

The choice of which grasses to grow depends on climate. Not all grasses thrive in areas of the country that are too hot or dry, and some do not do well in cool northern climates. On some farms, having some fields in warm-season grass and other areas growing cool-season grass can increase pasture productivity over a longer growing season. Cool-season grasses can be grazed when they are more productive in the spring and fall, and warm-season grasses during midsummer (see figure 3.4).

Warm-season grasses can produce highly palatable pasture when managed correctly, and they are at their most productive when cool-season grasses have gone dormant due to heat and drought. This makes them a good species to fill in the midsummer slump in pasture production (discussed earlier in this chapter). Some warm-season grasses are not winter-hardy, and will grow as a perennial only in southern regions. However, some warm-season plants such as switchgrass will persist in some northern regions, while others like millet or sorghum sudan grass will do well as annual summer crops.

The perennial warm-season grasses are often slow to establish; some species may take more than a year to become established well enough to graze. Allowing them to grow to 16 to 20 inches tall before grazing will help maintain plant health and vigor. It's important to leave a tall post-grazing residual compared with most cool-season grasses. Warm-season grasses should be left with at least 6 to 8 inches of growth in the fall before they go dormant for winter.

Prescribed burning is often recommended to maintain stands of some warm-season grasses, but with good grazing management this should not be necessary. Burning is done to remove excess litter and woody weed species. However, grazing with a higher stocking density will also remove litter and weeds and will return the organic matter and carbon to the soil instead of putting more into the air.

Native warm-season grasses once covered the Canadian and American prairies where bison roamed in vast herds. When cattle grazing began on the prairies, though, poor grazing practices and tillage led to the replacement of native species with other grasses. The native warm-season species include switchgrass, big bluestem, and Indian grass.

Legumes

Legumes are another important group of plants in pastures because they help build soil fertility and provide highly digestible, high-protein forage for livestock. Legumes have a symbiotic relationship with microorganisms called rhizobacteria that live in their roots. This relationship makes it possible to "fix" nitrogen from the air so that the plants can use it for growth. In exchange, the legumes provide carbohydrate energy to the rhizobacteria. The nitrogen collected in this mutually beneficial process also becomes available in the soil for other pasture plants such as grasses to use. Because of this, including legumes in a pasture mix can decrease or eliminate the need to apply nitrogen fertilizer. Legumes mixed with grasses also offer ideal forage quality for livestock. When planting legumes, it may be a good idea to include an inoculant of the right species of rhizobacteria with the seed.

As with grasses, regional differences in climate will determine which legumes will grow best on a specific farm. Chapter 6 goes into more detail on which plants are better adapted for hot, dry climates or cooler northern areas.

Legumes grow differently from grasses in several ways. In the spring, legume stems begin to grow in length immediately; there is no shift from a vegetative to an elongation stage as in grasses. Legumes have many potential regrowth points and can produce flowers on either side branches or main stems.

Within the legumes, some species are perennial and some annual. Some are very tall upright plants with deep taproots, while others are low growing. For example, alfalfa has an upright growth habit, but white clover has a horizontal growth habit, as shown in figures 3.17 and 3.18. Red clover and bird's-foot trefoil have intermediate growth habits.

Alfalfa plants are perennial and have a crown above the soil surface, so it may be damaged by trampling, which is why it may not persist under high-density grazing. Also, alfalfa will fare better when it is not grazed too

Figure 3.16. Red clover stems grow vertically, lifting the leaves, flowers, and growing points up in the pasture plant canopy.

Figure 3.17. The stolons of this white clover plant, which has a horizontal growth habit, extend across the soil surface. This keeps the growing point on the stem close to the ground, while the leaves and flowers grow up into the pasture canopy.

Figure 3.18. This lonely alfalfa plant is in a newly seeded field that suffered a lot of winterkill the previous winter. It's easy to observe how the crown and growing points of the plants are raised up above the soil surface.

short, because short grazing can allow livestock to actually consume the crowns, where many of the growing points of alfalfa plants are situated.

White clover, also a perennial, is one of the best grazing-adapted plants, due to its low-growing, spreading growth habit. However, it is also sensitive to shading and will disappear in overgrown pastures or pastures that are frequently left to grow up tall. White clover has relatively shallow roots, so it is also more likely to become less productive during midsummer's hot, dry weather than deeper rooted alfalfa or red clover. For this reason, white clover can show significant seasonal variations in growth.

Following defoliation, white clover leaves will rapidly regrow from the horizontal stolons. Over time those leaves will age, and other plants such as grasses and broad-leafed forbs will also grow up into the sunlight. When new clover leaves form and emerge under the pasture plant canopy, they will be in less light due to shading. The clover plant "knows" this, however, so it responds by elongating this new bunch of leaf petioles to lift the new leaves up into the light. This

adaptation is one of the reasons white clover does so well in pastures.

Clover has horizontal instead of vertical leaves as grasses have. This adds more density to the pasture canopy so it can capture as much sunlight as possible.

Clover exhibits phototrophic movement — movement that causes stems and leaves to orient themselves facing the sun. This is a growth advantage because the leaves then intercept more light throughout the day. This movement comes about when the plants elongate cells in the stems only on the side farthest from the light. This has the effect of bending or moving the stems and the leaves toward the sun.

White clover also responds to shade and sun on its stolons by either growing more branches or elongating. In densely shaded areas and clumps of grass, clover stolons (horizontal stems) tend to grow linearly. But when they reach a sunlit gap in the pasture, they will branch profusely to colonize the open area. This adaptation is one of the reasons that white clover spreads naturally in many pastures.[8]

In addition to the perennial legumes discussed (white clover, red clover, alfalfa), there are also several annual

Figure 3.19. On the Boomhower farm in northern Vermont, white clover has spread naturally throughout this pasture to create a dense canopy of leaves. This is providing high-quality forage to their organic dairy herd and capturing sunlight efficiently.

A DIVERSE PASTURE MIX

The types of plants that will grow naturally or do well when planted as part of a pasture mix will vary greatly depending on the local climate of the farm. The legumes in the mix provide highly digestible forage to the livestock, and nitrogen to the soil. Grasses provide fiber and energy. Forbs provide a mix of protein, energy, and other benefits. Taken as a whole, this diverse mixed pasture creates a denser canopy, which can capture sunlight and maximize photosynthesis.

legume species. Annual legumes such as peas and beans are generally grown as a grain crop, but can also be grown as a silage crop, as green manure, or for grazing. They are a high-protein feed and, when used as an annual pasture, are usually grown as part of a mixture with other plants. Growing annual legumes with upright annual grasses such as oats will provide support so the vining legumes stay upright. Mixing in grasses will also provide a broader range of forage nutrition. Strip grazing is one of the best ways to graze these annual mixtures, because it is easier to limit daily intake of the annual pasture and prevent trampling of ungrazed forage. (Strip grazing is explained in chapter 5, and in The Art of Good Grazing at the end of chapter 4.)

When grazing perennial and annual legumes, there can be a risk of bloat. Refer to chapter 12 for more information on bloat.

FORBS

Whether they are intentionally planted or not, there will probably be some forbs in any pasture. The word *forb* is generally used to describe plants that aren't grasses or legumes, though the official definition is "a flowering plant other than grass."

Some forbs may be weeds, while others are potentially productive pasture species. Some of these are annuals, which must be replanted or reseed themselves each season, and others are perennials that may persist as part of the pasture mix for years. Beneficial forage pasture forbs include forage chicory, dandelion, plantain, and forage

Figure 3.20. Dandelion is part of the diverse mix of plants in this pasture. With its diversity of forbs, grasses, and legumes, this pasture can grow during a wide range of weather conditions, maintain high plant density to capture sunlight efficiently, and provide high-quality forage to the dairy herd.

Figure 3.21. This newly seeded diverse pasture mix includes perennial grasses, plantain, legumes, and chicory. Plant density is low, which is common in a new seeding. Under good management the density should increase over time.

brassicas such as kale or turnip, which can play a useful role in a diverse pasture mix.

Many plants commonly considered weeds have interesting nutritional or medicinal uses. Dandelion, for example, is high in minerals and has some diuretic properties. Chicory is high in condensed tannins, which can help with management of internal parasites. In his 1955 book *Fertility Pastures*, Newman Turner wrote about "herbal leys." These diverse pasture mixtures include several grass and legume species along with forbs including chicory, plantain, burnett (*Sanguisorba minor*), and other species.

Browse

Browse plants are often woody brush or tree species that are *not* grazing-adapted. They are commonly found in hedgerows, overgrown pastures, and forest understories. They include deciduous trees such as birch and poplar as well as shrubs such as blackberry, raspberry, spirea, and goldenrod.

Not all woody shrubs and trees will make good browse due to palatability or toxicity issues. However, for some small ruminant farms, managing to include some browse in the grazing rotation may be desirable. This may include managed grazing of hedgerows, woody plants, or brush. Many of these brushy or woody plant species, as well as some of the "regular" pasture species such as chicory and bird's-foot trefoil, can serve as a source of condensed tannins, which can help with livestock parasite management. Additionally, the young leaves of these woody plants are highly digestible and nutritious.

If you make the decision to include browse in the pasture system, keep in mind that managing a pasture to maintain a significant amount of woody browse species is challenging, and may not always be practical or economic. If grazed too hard or too often, most woody browse species will die, and the pasture will convert to species such as grasses and clovers that are better adapted to grazing. To manage to maintain browse, graze the woody plants lightly. In addition, giving woody species a much longer regrowth and recovery period is often necessary. Some farms accomplish this by grazing these woody browse areas only lightly and only every other year. Another option is to maintain browse hedgerows, which can be grazed lightly on occasion. In brushy areas where the browse is being killed by grazing, planting some forbs such as chicory or plantain, which are better adapted to grazing than the woody browse plants, in the pastures can provide a diversity of plants as the brushy species die out under grazing pressure.

Annual Pasture Plants

Annual pasture species may also be a useful part of a grazing system. Annual plants can be particularly helpful in extending the grazing season into the fall or through the midsummer heat when perennials are growing slowly or have gone dormant. Annuals for grazing include both warm- and cool-season grasses, legumes, and some forb species. Some warm-season plants will grow as perennials in warmer climates, but in colder northern areas these may only grow as an annual crop.

Because annual pastures are replanted each year, they can be managed differently than perennial pastures. Perennial pastures, just like a hay field, must be managed carefully so that overgrazing does not damage the plants, which will reduce the next year's yield. When grazing an annual crop such as millet or oats, there may be multiple grazings in one season, but the goal can be maximizing livestock feed intake instead of trying to balance feed intake with careful long-term perennial plant management. Highly productive annuals that can be grazed include millet, sorghum sudan grass, oats, annual ryegrass, and legumes such as peas and annual brassica crops.

In addition to these intentionally planted annuals, some species that reappear on their own in pastures each year are annuals as well. These plants reseed themselves naturally, and include both productive palatable species and some that are considered weeds.

Caution: There are risks associated with grazing certain annual crops. For example, prussic acid poisoning can result when grazing sorghum sudan grass. To prevent problems, avoid grazing sorghum sudan during or right after a frost, and be sure plant height is at least 18 to 24 inches before grazing. (See chapter 12 for more detail.) Millet, which can be grazed at 14 to 24 inches of height, does not cause prussic acid poisoning. These annual warm-season plants are commonly used in midsummer when perennial pasture growth slows.

Forage brassicas such as turnips, rapeseed, and kale are most commonly grazed in fall or late summer. They maintain quality well into freezing temperatures and can be used to extend the grazing season on some farms. Health problems can occur if brassica grazing is not done correctly. These can include bloat, atypical pneumonia, nitrate poisoning, and hypothyroidism. Following the appropriate management practices (see chapter 12) should prevent these health disorders. First, introduce animals to brassica pastures slowly (over three or four days); avoid sudden changes from regular pasture to lush brassica pastures. Second, brassica crops should not be the only pasture or forage in the ration. Supplement with dry hay, silage, or "regular" pasture while grazing brassicas.

For farms with the right soils and access to equipment, growing annual pasture crops can be a strategy to assure a regular supply of high-quality farm-grown feeds. For other farms, it is clearly best to keep the focus on producing perennial forages. Each farm will have to assess the feasibility of growing annual crops according to their own goals and unique farm situation. Annual tillage and seeding cost money and require significant labor. In contrast, well-managed perennial grass–legume pasture provides many benefits in building and maintaining soil fertility with lower annual expense and labor. With a

Figure 3.22. At the Beidler Farm in central Vermont, the dairy herd is being strip grazed in millet.

good grazing management system, many farms may be able to use perennials only and avoid the additional labor and cost of growing annuals.

The Art of Good Grazing

Pasture and Browse for Dairy Goats

Does' Leap goat dairy, located in northern Vermont, makes fresh and aged organic goat's-milk cheeses with milk from the farm. Goats are natural browsing animals, so owners George VanVlaanderen and Kristen Doolin manage the farm so that the goats have access to both perennial pasture and browse. The perennial pastures are a mixture of cool-season grasses, legumes, and forbs. The browse includes a variety of small trees, brush, and brambles. The goats thrive on this diverse mixture of plants, and also benefit from the daily walks to and from the pastures.

The goats get a fresh pasture twice a day, after each milking. Then each pasture is allowed to regrow for several weeks or in some cases for months, depending on the time of year and plant species. The pastures that are primarily grasses and legumes regrow more quickly, while the pastures where they are trying to maintain more browse often need significantly longer recovery periods.

The goats are also fed a small amount of grain during milking. The herd is somewhat seasonal; most goats are dried off for winter, and kidding is timed so the does' most intense nutritional requirements occur when pasture quality is at its best.

Because most woody browse species are not well adapted to frequent or severe defoliation, over time many of the brushy pastures on the farm have naturally converted over to mostly grazing-adapted plants. This has improved the forage quality and productivity, but

Figure 3.23. Goats are naturally good at browsing, and their mouth and lips are flexible and strong. This allows them to selectively eat leaves of bushes, trees, and other browse plants while avoiding thorns.

Figure 3.24. This pasture at Does' Leap has a hillside area of browse as well as an area of grazing-adapted perennial cool-season grasses and legumes.

George and Kristen feel that for goats it is particularly important to maintain some browse in the pastures. To manage the browse so that it isn't killed by grazing, they use larger paddock sizes to lower the stocking density in browse pastures. This assures that plants retain some foliage; some plants may not be grazed at all during some grazing cycles. Despite Kristen and George's use of lower stocking densities and longer recovery periods to try to maintain some browse species, many of the areas on the farm that were primarily brush/browse plants continue to gradually convert to grasses and legumes.

George and Kristen find there are many benefits from having a wide diversity of pasture sizes and types of plant species. Larger brushy pastures with rocks and other topography the goats can climb on allow the flock to engage in their natural behaviors. In addition, the leaves of many of these woody and brush species also contain tannins, which can offer some help with the management of internal parasites. An additional benefit of browse is that it encourages the goats to graze plant parts that are higher above ground level, and thus they are less likely to become infected with internal parasites.

Kristen and George's flock is certified organic, and without their careful management of the flock and the diversity of pastures and plants on the farm, it would be much more challenging for them to control parasites without the use of synthetic dewormers. Refer to chapter 12 to learn more about the benefits of browse in parasite prevention.

CHAPTER 4

Pasture Soil Health

Caring for the pasture plants is inextricably inter-related with care of the soils in which those plants grow. Without a healthy soil, there is no chance pasture plants can survive and be part of a thriving pasture ecosystem.

Newman Turner, author of *Fertility Pastures*, wrote, "Fertility in the soil means to me the ability to produce abundant yields of crops which are disease-free, the seeds of which have the capacity to reproduce healthy crops and to transmit health, productive ability and fertility to the animal and human consumer."[1] Healthy, fertile pasture soils create a living foundation in which the highest-quality pasture plants can thrive and produce an abundance of forage. These high-quality forages allow livestock to reach their potential as healthy parts of the whole farm system. In return, cattle will produce manure, which if well managed is central to the healthy cycling of nutrients on the farm. Forages grown in these healthy soils are also converted by cows, goats, and sheep into milk and meat, which support human health and provide income for the farm.

In a healthy living soil, pasture plant roots will be able to anchor the plants and take in water and minerals to support plant growth. Soil should be full of a diversity of living organisms, from earthworms and dung beetles to bacteria and fungi. These organisms play an essential role in cycling nutrients and energy from soil, air, and sun to plant, animal, and manure and back to the soil. Soils must also have the physical structure to hold both enough water and nutrients that plants need to grow. For best plant growth, it's important that soils offer the right amounts of essential nutrients and also have a favorable balance of air and water.

Soils vary widely from region to region and can have very different physical, biological, and chemical characteristics. Variations in soil quality and health are due to many factors, including:

- The parent material that the soils developed from.
- The local climate that the soil developed in.
- Slope and drainage.
- The plant and animal life on and in the soil.
- Past soil management practices.
- Past applications of fertility amendments.

Poor soil quality limits what types of plants can grow and how productive they can be. Better-quality soils allow for more productive pasture growth.

What Are Healthy Soils?

The essential elements and characteristics of a healthy organic soil are soil life, organic matter, minerals, air, water, and the soil's physical properties. However, an array of physical, biological, and chemical factors contribute to soil health. Jerry Brunetti wrote: "Soil truly is a very complex ecosystem made up of ecosystems that know one another. The animate and inanimate are intertwined in a dance of synergy in which the whole is much greater than the sum of its parts."[2] The healthiest soils are the ones that have the right amount and balance of fertility, high organic matter, adequate drainage, and optimal soil structure. Healthy soils are teeming with life and biological activity.

With improved soil health, preferred pasture plant species will grow better and (with good management) should be able to compete well with weeds. The total amount of pasture dry matter per acre should increase, and pasture nutritional quality should also be better. Due to this high forage yield, the farm can support more healthy livestock and produce more meat and milk with less purchased supplemental feed.

On a grass-based livestock farm, the availability of manure produced on-farm, and the fact that much (or

Figure 4.1. Pasture plant roots grow vigorously in a healthy living soil. These larger, healthy root systems hold the soil and plants in place, which prevents erosion. More healthy roots also allow easier uptake of nutrients and water so plants are less susceptible to drought.

all) of the farm is in permanent perennial sod, creates both challenges and opportunities in the development of a sustainable approach to soil fertility. There is no single recipe to create a soil that is healthy biologically, physically, and chemically. Each farmer must creatively use the practices available within the generally understood guidelines of good management and apply an understanding of what healthy soils are. Find a knowledgeable soils consultant or a mentor, choose some appropriate methods of monitoring soil health, and study the basic principles of soil fertility. You'll also find listings of books and other resources about soil in appendix E.

THE LIVING SOIL

When you feed a cow well, you are feeding her rumen microbes. And when you feed the soil properly, you are feeding soil microbes. Soil microbes make nutrients available to the plants in much the same way microbes in the digestive system of a cow make nutrients available for her growth and production. Keeping the microbes well fed, housed, and cared for in both soil and livestock will generate the most meat and milk possible via contented cattle eating high-quality plants. (Chapter 7 goes into more detail on rumen microbes.)

In the rhizosphere (the microbe-rich zone in the soil around the root hairs), there is an incredible diversity and abundance of organisms. Jerry Brunetti wrote, "Soils contain an estimated 2–3 million species of bacteria and an estimated 1.5 million species of fungi, and only 2–5 percent have been described or named!"[3] In addition to bacteria and fungi, the soil is also home to other microorganisms including algae, lichens, nematodes, and protozoans. Larger organisms in the soil include mites, earthworms, spiders, dung beetles, ants, grubs, spiders, voles, and gophers. Some of these organisms play a role in decomposition, and some move materials around in the soil and create tunnels that aerate the soil. Some are predatory on other soil organisms.

Beneficial soil microorganisms play an essential role in making nutrients available to plants and even help to protect the plants from disease. In return, the plants then provide these organisms with energy. Most of us are familiar with the nitrogen-fixing organisms associated with clover and other legumes. They are just one example of the critical role that beneficial soil organisms play in overall soil health and productivity. Earthworms and soil fungi also produce sticky substances that help hold together soil particles so they can resist erosion and hold more water and the right amount of air (this is explained in more detail below).

In order to thrive, soil organisms need good drainage, adequate moisture and air, the right amounts and balance of nutrients, and the timely addition of organic matter. Well-managed perennial pastures create an ideal habitat for beneficial soil organisms. Crop rotations that include sod crops are particularly helpful in restoring beneficial organisms to tilled soils. The addition of compost can also increase soil biological activity. One easy measure of the health of soil biological activity is how fast manure breaks down in a pasture. Within minutes of landing, manure should already have dung beetles and flies on it. Microorganisms will be colonizing it, too. Soon after that, the organisms will aerate the pile and make it possible for other decomposing organisms to join in the process. On farms with excellent biological activity, the manure may no longer be visible by the end of the recovery period following a grazing.

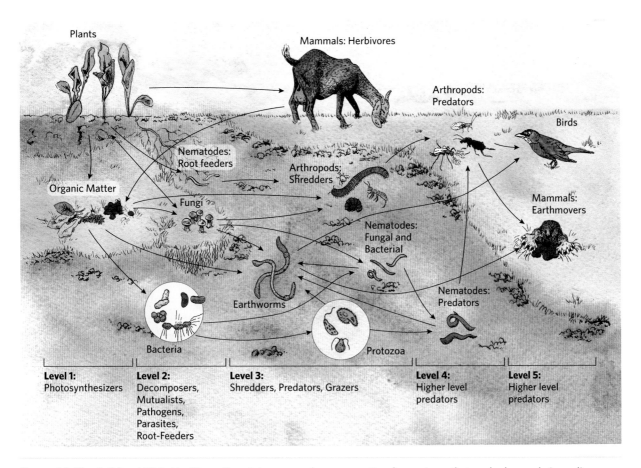

Level 1:
Photosynthesizers

Level 2:
Decomposers,
Mutualists,
Pathogens,
Parasites,
Root-Feeders

Level 3:
Shredders, Predators, Grazers

Level 4:
Higher level
predators

Level 5:
Higher level
predators

Figure 4.2. The Soil Food Web: Healthy soil contains a complex community of organisms that each play a role in cycling nutrients through the pasture ecosystem. Illustration by Angela Boyle, angelaboyle.flyingdodostudio.com.

ORGANIC MATTER

The amount of organic matter in the soil affects almost every other aspect of soil, and monitoring organic matter content is a central aspect of soil management. Fred Magdoff and Harold van Es wrote: "Good organic matter management is . . . the very foundation for a more sustainable and thriving agriculture."[4] Organic matter includes living organisms, fresh residues, and decomposed residues. Magdoff and van Es described these three aspects of organic matter as the living, the dead, and the very dead. The living organisms were briefly discussed in the previous section, but the residue in various stages of decomposition is also an essential part of what makes up the soil organic matter.

Organic matter is made up of plant residue, manure, dead microorganisms, earthworms, and plant roots. This is what makes up the food supply for many of the soil organisms. Organic matter also includes humus, which generally refers to the fully decomposed (very dead) part of soil organic matter. Humus is able to hold essential nutrients, which can then be slowly released to plants. It also reduces soil compaction and improves water retention and drainage problems.

By increasing soil organic matter, the farmer is feeding soil organisms, building stable soil aggregates, and providing nutrients. Organic matter improves water-holding capacity, drainage, porosity, and aeration. Organic matter helps bring balance to all soil types.

Soils high in organic matter have good cation exchange capacity (CEC). (Cations are positively charged nutrients such as calcium, magnesium, and potassium.) CEC is the amount of negative charge in the soil that makes it possible for soil particles to hold on to positively charged cations. This means that high-organic-matter soils can hold more magnesium, calcium, and potassium, which are essential for plant growth. High-organic-matter soils with high biological activity will also have more availability of other nutrients, including phosphorus and nitrogen.

"By a fertile soil is meant that nature's law of return has been faithfully applied. So that it contains an adequate amount of freshly prepared humus made in the form of compost from both vegetable and animal wastes," wrote Sir Albert Howard in the introduction to *Pay Dirt*.[5]

Methods of increasing soil organic matter and creating and maintaining healthy soils include:

- Rotation of crops so that tilled land is returned regularly to sod (grass/legume).
- Good management of permanent perennial pastures.
- Regular additions of compost.
- Minimization of tillage.
- Prevention of soil compaction.
- Covering bare soil whenever possible by undersowing into annual crops such as corn, and by using green manures, mulch, and cover crops whenever possible.
- Prevention of soil erosion.
- Minimization of highly soluble fertilizers or synthetic pesticides.

MINERALS AND PASTURE NUTRIENT CYCLING

Managing the nutrients cycling through a pasture system is very different from doing so in a system where crops are grown and harvested and fed to the livestock in confinement. In a pasture system, nutrients cycle from the plants as they are consumed into the livestock, then back onto the pasture soils in the form of urine and manure. A relatively small amount of the nitrogen, phosphorus, and potassium in the plants that the livestock graze is removed from the farm (in the form of meat and milk). With careful management, most nutrients can be cycled back into the soils and plants. Refer to Manure and Soil Health later in this chapter to learn more about how

grazing management can optimize spatial distribution of manure in pastures.

In a diverse pasture being managed well with a group of livestock, nutrients for plant growth come from many sources. Livestock provide manure as they graze in the pasture, and the action of their hooves tramples plant residue into the soil. Legumes fix nitrogen, and soil organisms move nutrients around in the soil, and ultimately their decomposing bodies contribute nutrients, too.

Either a deficiency or an excess of any mineral can lead to problems with plant growth, resulting in lower forage quality and less-than-ideal nutrition for livestock. Many soil nutrients can be added in the form of compost or manure, or from legumes. However, if a specific mineral is deficient and on-farm compost or manure isn't able to bring the soil into balance, it may be necessary to purchase off-farm amendments. Refer to the troubleshooting table in appendix A to see some common symptoms and solutions for soil deficiencies and imbalances.

Soil tests are an important tool to determine if soil minerals are deficient or out of balance. Soil testing is essential when adding minerals to avoid overfertilizing or creating new imbalances. See Measuring and Assessing Soil Health later in this chapter for more on this topic.

AIR, WATER, AND PHYSICAL PROPERTIES OF SOIL

In order to allow healthy plant growth, soils need to be able to hold enough water to get through dry spells and contain enough air so they don't become anaerobic. They should be able to resist erosion by wind or water, absorb water quickly during a rain, hold nutrients, and sustain an array of soil organisms and plant life.

The basic texture of soil relates to the mineral from which it is made (parent material) and its geological history. Pasture soils may be naturally rocky, sandy, silty, or clay-based. Data about local soil types and the parent materials can help with a better understanding of the pasture soils and what their potential is.

Soil structure, which shows up visually in a shovelful or a handful of soil, refers to how soil holds together in a clump (aggregation) or breaks apart. A soil with good structure and normal moisture conditions will crumble easily in your hand, but will hold together if squeezed.

Converting annually tilled and planted fields to a perennial grass legume mix can rapidly improve soil structure. The mass of fine roots from the grass plants helps in the formation of soil aggregates. These roots provide soil microorganisms and earthworms with food, and those creatures then make glue-like compounds such as glomalin, which helps bind soil particles together. These new soil particles will be more able to stay intact even in a heavy rain, so the soil remains porous, can hold water, and is less likely to blow or wash away.

Practices that improve soil structure include:

- Minimal reliance on tillage.
- Crop rotation that includes a sod.
- Good grazing management.
- Avoiding equipment use in fields when soil is too wet to prevent compaction.
- Any method that adds organic matter.

One of the many benefits of good grazing management, with short grazing periods and ample time for pasture plants (and roots) to regrow, is that over time this will create better soil structure. Agronomist and author Preston Sullivan wrote: "The best-aggregated soils are those that have been in long-term grass production."[6]

Some soils have naturally occurring physical problems that will limit pasture productivity no matter how much is spent on fertilizers. For example, a soil may be poorly drained due to a hardpan or impervious layer. The area at the bottom of a slope where water collects may also be poorly drained. This poor drainage will limit what types of pasture plant species can grow in the soil.

Wet Pastures

In poorly drained pastures, it may be possible to improve existing ditching or drainage tile. However, water-quality regulations may prohibit installation of new tile and ditching. Be sure to check with local NRCS or soil conservation staff before starting to drain a wet area.

If a pasture actually has standing water in it at some periods each year, this will severely limit what can grow there. Standing water, unlike flowing water, will rapidly

Figure 4.3. Sedges resemble grasses, but they are not grasses. These plants aren't very palatable or digestible, but they're able to grow in areas with saturated soils. The triangular stem of this plant helps identify it as a sedge.

deplete oxygen available to plant roots. The majority of the improved pasture forage species will not tolerate oxygen-depleted conditions, and they will be replaced by wetland species such as rushes and sedges. Sedges, rushes, and other species are well adapted to growing in fully saturated soils. These wetland areas will never become productive pasture areas, so effort and investment should instead be made in improving pastures with soils that have a more productive potential.

Wetlands are an important part of our farm ecosystems, because in high-rainfall events they provide necessary buffers that can prevent flooding damage. These areas also provide habitat for many beneficial plant and animal species, such as amphibians and birds. So some of the wet areas on the farm may best be "managed" by fencing cattle out of them permanently, or allowing only very limited short-duration grazing at certain times of the year. If the area is so wet that no high-quality forage grows, fencing livestock out is probably the best choice. However, if the area is only seasonally wet and contains some browse or other forages of value, then grazing may be an option. It's important to time the

Figure 4.4. Protecting wetland areas is an important way to build resilience into the farm landscape. These areas provide buffers, which can protect against floods during high rainfall events. On Tussock Sedge Farm (see chapter 6), many wetlands have been protected with fencing. Birdhouses on fence posts along the riparian edges increase biological diversity on the farm.

grazing so that livestock are not damaging wet, saturated soils. In addition, very short-duration grazing, also called flash grazing, is usually best to prevent damage to water quality or soils.

Droughty Pastures

Soils may be droughty due to the presence of stones, sand, and gravel, resulting in coarse texture. Low organic matter, poor soil structure, or a shallow soil due to bedrock or hardpan close to the surface can also cause droughtiness.

In pastures where the droughtiness is due to bedrock close to the soil surface, only limited progress is possible

UNDERSTANDING SOIL TYPES

A soils map for your area is generally available from your local Natural Resources Conservation Service office (NRCS). You can also find these online (see appendix E); there are even apps for your smartphone that tell you the soil type in the field you are standing in! These maps are from surveys or studies of soils that have been done throughout the United States and Canada. They show the distribution of different soil types and provide descriptions of the soil types and what their primary uses and limitations are. Studying soils maps and descriptions of farm soils can help you understand the potential productivity of each pasture.

to improve the productivity of the area. However, if low soil organic matter is the cause, there are many strategies to build more soil organic matter and improve water-holding capacity.

Erosion

Some soils are at a higher risk of soil erosion. This may be due to previous management, but may also be due to the slope and soil type. Allowing any soil to erode from our farms is unacceptable. Soils are the foundation of our farms and our food system, so every effort must be made to keep them where they should be — in our productive pastures.

When grazing erodible soils, make the extra effort to prevent the formation of animal trails that could allow water and soil to run downhill. Limit grazing to short durations, and be sure to maintain the pasture in a dense perennial sod. The more cover of vegetation and residue there is on the soil surface, the lower the chance that any soil can erode. In addition, efforts to improve soil structure and build organic matter will reduce risk of erosion.

Measuring and Assessing Soil Health

Each farmer develops his or her own way of assessing soil health and monitoring changes over time. This may include tracking changes in crop yields, recording changes in the types of plants growing in the pasture, or sending soil samples to a lab for testing of nutrient levels. (Appendix C includes a simple worksheet for pasture monitoring.) For an experienced farmer, it may also include a more intuitive ability to look at the soil, smell it, feel it, see how the plants are growing, and know whether it is improving or not. Dr. Hubert Karreman wrote, "If we dig up healthy soil, especially in the early morning or near dusk when dew is present, an invigorating burst of pure moist earth strikes us at a most visceral level, invigorating something quite instinctual within us."[7]

Soil Testing

Soil testing provides valuable information for making decisions such as when to invest in fertilizer, which type of fertilizer will be most helpful, and what types of pasture plants can be planted. It is essential to have a recent soil test before spending money on reseeding a pasture. Without a soil test, it is not easy to know which minerals are lacking, and those mineral imbalances may prevent the newly seeded plants from growing well.

Different soil testing labs use different mineral extraction methods, so it is best to choose a lab and stick with it instead of using a different lab each year. This will allow a more accurate evaluation of how soil fertility is changing over time in the pasture.

It is also important to use a lab familiar with the soils in your region. It's more likely that a local lab will use extraction methods well matched to the local soil types.

How to Take a Soil Test

Collecting a soil sample properly is crucial. You want to make sure it's truly representative of the whole pasture you're sampling. Once you've chosen a lab, read its soil testing kit instructions carefully. The instructions will tell you how large the aggregated soil sample should be and will often provide other helpful information on techniques for collecting individual samples. Samples can be taken at any time of the year. However, it is easiest to take the samples when the soils are thawed, moist but not too wet. Samples should never be taken right after the application of lime or fertilizer, as that will

contaminate them. Soils do change over the year due to plant activity and weather conditions, so it is best to take the samples at the same time each year. Fall is often a convenient season to sample, because it allows time to receive results and make decisions over the winter on the farm's nutrient management system for the following season.

It may not be necessary to take a separate sample from every individual pasture on the farm. If two or more pastures have the same soil type, are growing similarly, and are managed and fertilized the same way, you should be able to assemble an aggregate sample of those pastures and submit it for a single test. But if you have pastures that are growing quite differently or have been managed differently, it's a good idea to take samples separately. In any pasture, avoid taking samples from areas that are unusually wet and areas under trees where livestock congregate. Also avoid hedgerows, field edges, or other areas that are not typical of the pasture.

It's easiest to collect samples using a soil sampling auger or tube, but a shovel or trowel will serve the purpose if that is all that's available. Make sure that the equipment you use is clean. You may need to clear plant residue from the surface of the soil to make the sampling easier. If you use a shovel, the holes will be V-shaped. Take a thin slice of soil from the exposed surface of the hole at each sampling site. Using an auger rather than a shovel makes it easier to collect a more consistently sized soil core. As you collect individual samples, put them in a clean bucket. Collecting and mixing 20 to 30 of these cores or subsamples from randomly selected locations throughout the pasture results in a representative sample of the whole pasture.

The depth that the samples are taken from will vary somewhat depending on soil type, how deep the plant roots are, whether there is a physical obstruction such as bedrock, and what the past management has been. Some farmers who have permanent perennial pastures prefer to sample only the top 3 inches, while others sample to 6 inches or deeper.

As you take samples, observe root depth. This will provide useful information on how healthy and deep the pasture soils are. Knowing at what depth most of the pasture plant roots are growing can also help determine how deep the soil sample should be. The goal is to sample the region of the soil where roots are actively growing and absorbing nutrients.

Soil samples should be put in a clean pail so that nothing contaminates them. Potential sources of contamination include lime or fertilizer recently spread on the soil surface, or stray materials on the shovel or in the bucket. After you've collected 20 to 30 samples from one pasture, mix them gently and put some of the mixed soil into the plastic sample bag. The sample bag will generally hold a quart or less of soil, but this does vary with different labs.

The lab will provide a sample information sheet, which will ask for information on what crop is growing in the field and how it will be managed. You may need to specifically request a test for organic matter and trace minerals in addition to the usual macronutrient chemistry testing. If you want the lab to send you fertilizer recommendations that are limited to materials allowed on certified organic farms, be sure to specify that in the information sheet.

Particularly on a new farm, testing soil in most pastures every two to three years is advisable. Testing should generally be done more often when annual crops are being grown or fertilizer is being spread frequently.

Reading the Soil Test

Different labs report information in different orders and detail. In general, most lab reports provide information on the sample location, field name, and crop grown. The report will also usually list the pH, phosphorus, potassium, calcium, and magnesium levels. This section of a report usually also includes the cation exchange capacity. Some reports list the organic matter content of the soil and the trace mineral content including zinc, copper, boron, and manganese.

Amounts of the minerals may be listed as pounds per acre or parts per million (ppm). (To convert from ppm to pounds per acre, just multiply by 2.) Some measurements such as organic matter will be listed as a percentage. Many labs will also give some valuation of levels such as "low," "optimum," or "high." The soil tests usually provide recommendations of how much fertilizer to spread in pounds per acre or tons per acre.

MORE COMPREHENSIVE SOIL TESTS

Several labs and organizations now offer even more comprehensive soil testing, so that instead of just testing the soil pH and chemistry, many other measures are used. This includes measures of soil fungal and biological activity, plant root health, soil compaction, and many other interesting and useful ways to look at soils. Refer to appendix E for a listing of soil testing labs.

Soil pH is the measure of how acidic or alkaline the soil is. Soils under 7.0 are acidic and over 7.0 are alkaline or basic. The lower the number, the more acidic the soil is. Soil pH affects microorganism activity and the availability of many soil minerals. This is why maintaining an appropriate soil pH level will allow plants the best opportunity to grow well. Many of the grass-clover pasture mixtures do best with a pH of about 6.0 to 7.0. But some of the grasses will do well in soils with a pH of 5.5 or higher. Some pasture plants, such as alfalfa, are pickier and prefer a pH of 6.5 or higher. At soil pH of 5.5 or lower, many pasture plants will be less productive or may not persist over time.

Other Ways to Assess Soil Health

Formal soil testing provides invaluable information, but don't let it be your only method of learning about the health of your soil. Try some of these methods of assessing soil quality, too.

Observe the soil. Look at soil color, structure, and texture. Smell the soil.

Walk through your pastures, looking for areas that are droughty or poorly drained. Check for areas of soil compaction or a hardpan below the surface. One low-tech way to do this is to try to push a fence post into the ground. How much effort does it take? Another option is to use a penetrometer, a tool shown in figure 4.5.

Observe the plants. Dig a hole in the soil to see how deep plant roots grow, and how this changes over time. Is there a pan that prevents roots from going deeper? Look for plants with symptoms of deficiencies (yellowing, purple color, or spottiness).

Keep some records of your observations over time, too, watching for changes in yield, growth patterns, or the types of plants growing in different areas. Appendix C has a simple monitoring worksheet you may want to use.

Monitor biological activity. Observe earthworm activity by counting the number of worms or wormholes per shovelful or looking for worm castings on the soil surface. How is this changing from year to year? Watch for where dead plant material sits on the soil surface, and where it is more quickly incorporated into the soil.

Monitoring the rate at which manure decomposes can help us understand how healthy the soil biological activity is. Keep track of how long it takes for manure to completely disappear from the pasture, and the variety of creatures that assist this process of nutrient cycling. Can you find a population of dung beetles?

Figure 4.5. A soil penetrometer is a tool used to measure the extent and depth of soil compaction.

Figure 4.6. This farm has an excellent diversity of organisms that help quickly break down manure in pastures. This manure landed here less than two weeks ago and is rapidly disintegrating and being incorporated into the soil.

Figure 4.7. Even after a heavy rain, the water running off this well-managed pasture is still clear rather than cloudy. Good plant cover of the soil and excellent soil structure are preventing erosion in this pasture.

Watch the water. Spend some time in your pastures on a rainy day. Watch to see how the soils absorb water and how long they hold moisture. Water flowing off the pastures should not be cloudy or muddy, which is a sign of soil erosion.

Monitor for good soil structure so that rainfall is not landing on a crusted or impervious layer. Rain should soak in rapidly, and soils should be able to hold water but not become waterlogged.

Addressing Imbalances and Deficiencies

Once you have test results indicating what is too low, too high, or out of balance in the soil, you can make decisions about what soil amendments or management strategies to use. Appendix A also describes some of the common symptoms of soil deficiencies and imbalances.

On an organic farm, sources of fertility are limited to natural lime, rock powders, manure, compost, and other nonsynthetic fertilizers. For non-organic farms, there are also many synthetic fertilizers available to choose from.

pH. Adding lime or some other calcium source is the most common way to raise pH in pasture soils. Lime takes some time to raise the pH, so plan ahead and don't expect immediate results when starting with an acidic soil. The soil test report will supply a recommendation of tons of lime per acre to spread. However, if the soils are quite acidic and the recommendation is for several tons of lime, it is best not to spread it all at once. Instead, divide the recommended amount into several smaller applications over several years.

There are several types of liming materials, including some that contain magnesium as well as calcium. Some types of lime have been processed or heat-treated. Lime is not the only material that can raise pH, however. In some regions, less expensive materials such as wood ash may be used.

Nitrogen (N). Nitrogen deficiency will significantly limit pasture growth. However, it is available for "free" from pasture legumes and from manure spread by livestock as they graze. Spreading stored solid or liquid manure also provides nitrogen. Maximizing nutrients from manure is discussed later in this chapter.

For organic farms, off-farm sources of nitrogen fertilizer are limited and costly. For non-organic farms, synthetic sources of nitrogen may be useful if pastures have high nitrogen needs, low legume content, and inadequate manure sources.

Phosphorus (P). Phosphorus availability in the soil is affected by pH and is also very dependent on soil biological activity. Improving the health of soil life can help prevent phosphorus deficiencies. Phosphorus can also be provided by manure, rock phosphate (natural source), or synthetic sources.

Potassium (K). Potassium is very important for legume health. It is available from synthetic sources, from manure, or from natural sources such as greensand (rock dust) and sulfate of potash magnesia (also known as Sul-Po-Mag). It is important not to overapply potassium to pastures, because overly high potassium levels in forages can cause metabolic problems in grazing livestock.

Magnesium (Mg). Magnesium is important for plant growth and health, but if levels get too high it can cause problems for both soils and plants. Some lime sources contain magnesium, so these can be useful if soil levels of magnesium need to be raised. However, if soil magnesium levels are already optimal but calcium is still needed, it's advisable to seek out a type of lime that does not contain magnesium. Magnesium is also available in several synthetic fertilizer forms and in some natural sources such as sulfate of potash magnesia.

Trace minerals. Trace minerals or micronutrients are essential for soil microorganism health and plant health. Some of the most important micronutrients in pasture soils are boron, copper, zinc, manganese, molybdenum, and cobalt. These nutrients are needed only in very small amounts. Micronutrient testing should always be done when soils are tested, and if levels are low they should be added if possible.

If levels of some micronutrients are too high, these can also be highly toxic to plant and soil life. Before spreading any trace minerals, it is very important to test the soils and be sure that the correct amount is being spread.

The rate of trace minerals to apply per acre ranges from one to a few pounds. For this reason, it can work well to mix in trace minerals with other fertilizers so they can be spread evenly at a low rate on the pasture.

Manure and Soil Health

The manure produced on a grass-based livestock farm is just as important a resource as the meat or milk. Manure's role in pasture productivity and fertility is so critical that scientists write scores of articles on cattle defecation rates, quantities, and distribution on pasture. One cow can produce over 50 pounds of manure each day, which adds up to 9,000 pounds per 180-day grazing season or 18,000 pounds per year.[8] Such mountains of manure can be a great benefit to the pasture and soil health if managed well.

But if poorly managed, manure in pastures can cause water-quality concerns, uneven plant growth, rejected forage, and other problems. If the pastures are poorly managed and livestock have continuous access to large areas, their manure will be unevenly distributed. Manure will build up so much in some areas that it will block sunlight from reaching the plants. Grazing livestock will reject plants that grow in and around the manure, causing increasingly uneven grazing and plant growth.

Spatial Distribution of Manure in Pastures

Most of the nutrients livestock consume as forage end up being excreted. Research shows that dairy cows retain more N, P, and K in their meat and milk than beef animals do. Thus, a dairy herd may remove more nutrients from a farm's nutrient cycle than a beef herd. However, the nutrients in the purchased grain fed to dairy cows can result in a net import of some nutrients to the farm.[9]

Managing manure from both grazing and confined livestock is critical to the nutrient cycle of the farm. To do this well, even distribution of manure throughout the

Figure 4.8. Livestock such as the beef herd in this photo have a habit of bunching up in the shade even when it isn't hot. This can result in areas like this where plants have been killed by trampling and excessive manure.

Figure 4.9. An ideally designed shade pasture has evenly distributed trees that allow enough sunlight to penetrate for forage growth, but also provide enough shade for livestock.

pasture is important. The type of grazing system and the stocking density will determine whether the animals spread the manure around evenly, or instead spend much of their time dropping their manure under a pasture's few shade trees or near the water tank.

During a single grazing season, manure from grazing livestock will land on anywhere from 10 to 35 percent of the surface area of a pasture.[10] These areas of urine or feces also influence the area surrounding the landing spots, so under high stocking rates with very even manure distribution, more than half of the pasture may receive some fertility from manure. Even in well-managed pastures, however, not all the nutrients in the manure make it into the soil and plant roots. Some nutrients get lost due to volatilization and leaching, so additional soil fertility inputs may be needed. Soil testing is the best way to determine whether manure spread by the animals is enough to maintain soil fertility.

If livestock are given free range of a pasture (low stocking density) for a period of days or weeks, they will distribute manure very unevenly. The location of shade and water and the topography will influence where manure is deposited. If low-stock-density continuous grazing continues for years, there will be a steady transfer of nutrients from parts of the pastures where livestock choose to graze to the spots where they drink and rest. This can cause water-quality problems if nutrients build up in soils near streams or rivers. It will also cause a decline in pasture productivity and quality because soil in the areas where the livestock graze won't have sufficient manure returned to it.

"Dragging" pastures with a chain harrow or some other device after grazing is a common pasture practice intended to spread out the manure. Dragging can be helpful in some situations in continuously grazed pastures where manure distribution is uneven. However, it can also damage pasture plants that have horizontal stems at the soil surface, such as white clover. A better solution to poor manure distribution is to make smaller paddocks with a shorter period of occupation so the animals spread their own manure more evenly!

Location and distance of water also have a significant influence on manure distribution. Pastures where livestock have to walk down a lane or back through previously grazed pasture tend to have significant disparities in manure distribution. Livestock will graze in one area but deposit manure either at the water source or in the lane on the way to water. Setting up smaller

Figure 4.10. The best solution to assure even manure distribution in pastures is smaller paddocks, shorter periods of occupation, and grazing with a higher stocking density. The manure in this paddock is evenly distributed after being grazed by the dairy herd for part of one day.

GET LIVESTOCK TO SPREAD THEIR OWN MANURE

Here's a summary of strategies to improve distribution of manure in pastures:

- Use a higher stocking density, smaller paddocks, and shorter grazing periods.
- Use a portable feeder for minerals and salt, placing it in different locations in each new grazing period.
- Provide water in each paddock instead of having livestock walk down a lane to water. Portable water tubs connected to water pipes will allow the tub to be located in a different place in each new grazing period.
- Provide shade only when needed.
- If you're providing supplemental feed in the pasture, feed it in different locations in each new grazing period.

paddocks with water available in each results in more even distribution.

Grazing livestock also tend to deposit more manure in the areas where they rest or sleep, because they naturally defecate and urinate when they stand up. In rolling pastures, they may choose a particular hilltop as their "camping" spot, and over time this area may receive a lot of manure. The forages in these overfertilized areas will then be rejected; over time ungrazed forages and weed species will accumulate in these regular "camping" spots in pastures that are continuously grazed with a low stocking density.

Shade is necessary for livestock in some hot and humid weather conditions. However, shade can create significant manure nutrient transfer issues. If shade is always available to the livestock when they are grazing in a particular paddock, they will deposit much of the manure in the shade, instead of in the productive pasture areas where it is needed. This will happen both on cool cloudy days and on hot days, since livestock will often spend time in the shade out of habit. Shade should be made available to livestock when needed, but it may be

a detriment to soil fertility and pasture productivity to allow livestock access to the shade areas all the time.

If livestock are grazing endophyte-infected tall fescue pastures, keep in mind that cattle consuming the endophytes are more attracted to shade, and the gradient of manure in the shaded areas of those pastures may be more severe. (See the Fescues section in chapter 6 for an explanation of endophyte-infected grasses and chapter 12 for discussion of the health concerns.) Make sure that shade and other strategies for preventing heat stress in hot weather are part of your grazing plan, but make sure as well that habitual use of shade areas isn't causing manure-distribution problems. This is a good reason to have some dedicated shade pastures on the farm where livestock can graze during those hot, humid days when shade is physiologically necessary. Also note that when shade trees in a pasture are spread out, cattle will be less likely to bunch up in one particular location.

As noted in figure 4.10, the best solution to assure even manure distribution in pastures is an approach that combines smaller paddocks, shorter periods of occupation, and grazing with a higher stocking density. Additional solutions include providing water in each paddock, using a portable feeder for minerals and moving it regularly, and providing shade only when conditions require it.

SPREADING MANURE ON PASTURES

In addition to the manure the animals spread directly on the pasture, manure collected during the non-grazing season may also be spread on pastures. This may include liquid manure, solid manure, and composted manure.

On many farms, it will make more sense to spread this stored manure or compost on annual or perennial crops that are not receiving any manure directly from grazing livestock. However, spreading manure or compost on pastures may also be a great method to improve yields and quality. Manure can also be brought in from other farms to spread on crops and pastures. For certified organic farms, manure from other farms, even from non-organic farms, is acceptable under National Organic Program (NOP) regulations, so this can be a relatively low-cost solution if nitrogen is needed.

Different types of manures have different compositions. They also vary in how evenly they spread onto the pasture and how rapidly they will break down into the soil. If several types of off-farm manure are available in your area, it may be worth testing the nutrient content of the manure as well as the soil. Many labs that do soil tests also offer manure testing.

When spreading manure on pasture, it is also important to consider how soon the pasture will be grazed, and if the presence of the manure is going to result in more rejected forage. Manure that is either very fresh, poorly decomposed, or anaerobic is more likely to cause animals to reject forage. Aged well-rotted manure or composted manure will generally cause less rejection, particularly if it is spread with a small particle size so that the action of beneficial soil organisms and rainfall can incorporate it into the soil more rapidly.

PLANT AND ANIMAL RESPONSE TO MANURE

Pasture plants will generally show a short-term response to nitrogen in manure. This is particularly noticeable in nitrogen-deficient pastures, where there will be a lush, bright-green growth of grass in spots where urine has landed among the slightly yellow, nitrogen-deficient plants. The phosphorus and potassium in manure also provide fertility to the plants, but plant response to those nutrients is generally more long-term.

It is important not to overapply manure or other fertilizers to pastures. Overapplying one nutrient may cause soil imbalances and poor plant growth. It can also cause metabolic problems in the grazing animals.

Grazing livestock will naturally choose not to graze pasture plants that have manure on their foliage or are growing next to manure. This is sensible because that manure is a potential source of internal parasites. So around each area of manure in the short term there is likely to be an area of taller overgrown plants that livestock will avoid. While the nutrients from manure are beneficial, these areas of rejected forage can create some challenges, particularly for farmers new to grazing.

When manure lands on the ground, it creates a "zone of repugnance" where it lands and in its immediate surroundings. By using a higher stocking density, a farmer can force cattle to be less picky about what they eat, and thus the size of these rejected areas will be smaller in these circumstances. However, it isn't a good idea to force animals to eat everything in a pasture, both for the health of the plants (as discussed earlier) and also for the health of the cattle, because we don't want them consuming infective parasite larvae.

Different types of manure end up distributed differently in a pasture. Sheep and goat manure often come in the form of small pellets, which spread uniformly. Cows on very lush pasture may have such loose manure that it may spread out very thinly and break down rapidly. Cows consuming less digestible forage will form rounder and stiffer manure. As the biological activity of the soil changes, the amount of time it takes for manure to decompose and become incorporated into the soil changes, too.

When you're spreading manure or compost, there are some strategies you can use to decrease the amount of forage that will be rejected when livestock next graze that pasture. Spreading composted or at least partially composted manure will decrease rejection, as does spreading in the fall to allow manure time to break down before spring grazing. Spreading solid manure with

Figure 4.11. The small area of lush-looking grass around this manure is the "zone of repugnance" that cattle will avoid eating from if possible. These zones are often much larger areas of tall plants around fresh manure.

a type of spreader that breaks it up more finely (side slinger or auger-type spreader) results in a thinner layer of manure that will break down faster. There will be less rejected forage in a pasture if the manure has been rained on several times or irrigation has helped move it into the soil. Longer regrowth periods also allow more time for manure to decompose. In pastures where annual crops are being grown for grazing, the tillage process will incorporate manure into the soil. This can reduce nutrient loss to volatilization and also reduce rejected forage.

The biological activity in pasture and soils will also determine how long these manure areas remain and create rejected grazing areas. Pastures with a higher population of dung beetles, earthworms, and organisms that break down manure and move it into the soil will have fewer areas of manure and rejected forage. Dragging pastures should not be necessary as a way to spread out manure once the grazing is being done with the correct stocking density, number of paddocks, and length of grazing period.

The Art of Good Grazing

Grazing What Grows Best Locally

Elixir Farm is located in the Missouri Ozarks and is certified to both biodynamic and organic standards. This diversified farm produces grassfed beef as well as medicinal herbs, seeds, mushrooms, and produce.

The farm encompasses a wide variety of soil types, some of them stony and challenging to manage. Hot summer weather, particularly in dry conditions, creates additional constraints to growing many of the desirable cool-season grasses. In hot weather in this climate, cattle must have access to drinking water, and they also need some access to shade. Over time farm owner Lavinia (Vinnie) McKinney and her farm partner, Daniel Roth, have come up with creative ways to provide water and shade for the cattle and manage a very diverse mix of pasture plants. This has allowed them to extend the grazing season to 10 months or longer most years.

Vinnie and Daniel make a small amount of hay each year, which is fed to the cattle during winter months when the pastures are too icy or soils are too wet. This short winter hay feeding season actually has a benefit for the rest of the farm, because it allows the farmers to collect some manure and waste hay in a bedded pack, which is then made into compost to provide fertility for the annual vegetable crops.

Pasture plants on the farm include a mixture of cool-season grasses, warm-season grasses, legumes, and forbs. The primary cool-season grass that does well on the farm is tall fescue, since it is able to survive the summer heat and dry periods. This provides good fall and winter stockpiled pasture. One of the primary warm-season grasses on the farm is Johnsongrass. Over time Vinnie and Daniel have been able to reduce the amount of this grass by grazing it severely in the spring or early summer. This has allowed some other grasses and legumes to start growing in the pastures to add more diversity. Legumes include white and red clover, but there are also areas of the farm where sericea grows (see chapter 6 for more about this legume).

Over time, good grazing practices have resulted in huge improvements in soil health at Elixir Farm,

Figure 4.12. This healthy, happy cow is grazing a pasture that contains Johnsongrass. This warm-season grass is considered a weed on most farms, and can have some toxicity concerns for the cattle. On Elixir Farm, however, the use of carefully planned grazing and cattle management keeps cattle healthy and reduces the amount of this grass.

including a large increase in soil organic matter content. Vinnie and Daniel use a strip grazing method in most pastures. This allows them to provide the herd with fresh pasture once to several times a day by moving a fence forward into fresh pasture frequently.

By changing the size of the strip of new forage that the herd gets, or the frequency that the fence is moved forward, they are able to control the stocking density of the herd. The smaller the strip and the more often they move the fence, the higher the stock density they can use. A higher stock density creates more significant hoof impact and trampling. It also changes how much of the pasture the animals eat and how severely the herd grazes. Over time this trampling impact has reduced problem weeds and increased the density and productivity of the

Figure 4.13. The beef herd is being moved forward (to the right side in this photo) into a new strip of forage.

Figure 4.14. This photo shows the pasture after it has been grazed. The cows have grazed all the most nutritious leaves, and have trampled many (but not all) of the weeds and grass seed stems.

more desirable pasture plants. As part of the strip grazing system, a back fence is moved behind the herd every few days to prevent the animals from wandering back to parts of the pasture where plants may be starting to regrow.

To keep the cows and their calves eating the highest-quality plants in the pasture, Vinnie and Daniel have the herd graze only one-third to one-half of the plant material; the rest is left behind as standing or trampled residual. Because some leaves are left intact, the plants don't need to depend on stored energy reserves so much as they regrow. This is particularly important on the farm during drought conditions when plants are already experiencing some stress. Any plant stems that are trampled and left behind are also helping to build soil organic matter and improve fertility.

The cattle also benefit by not being forced to eat all the plants in the pasture as they rotate through. (Both tall fescue and Johnsongrass have some potential toxicity issues.) So by maintaining a diversity of species in

Figure 4.15. Over time with good management, even the most unpalatable weeds—such as prickly pear—have become less of a problem in the pastures of Elixir Farm.

the pasture *and* not forcing the cattle to eat everything, Vinnie and Daniel have been able to avoid the herd health problems that these grasses can cause.

The pastures all have areas of forest or hedgerows between them, and most of the pastures have been designed so that in hot weather cattle can be given access to an area with trees. During cooler weather, however, Vinnie and Daniel often fence the cattle away from the trees to make sure their manure and grazing impact is focused on the pasture plants.

— CHAPTER 5 —

Managing Pasture Plants

In *Essays Relating to Agriculture and Rural Affairs* (1777),[1] James Anderson of Scotland urged farmers to subdivide pastures into smaller paddocks, graze each one for a day, and then keep the animals out so the plants could regrow. Sounds like good grazing management! All this was long before the invention of electric fence, so Anderson had to use stone walls and a lot of labor to create paddocks. Anderson wrote:

> To obtain this constant supply of fresh grass, let us suppose that a farmer who has any extent of pasture ground, should have it divided into 15 or 20 divisions, nearly of equal value: and that, instead of allowing his beasts to roam indiscriminately through the whole at once, he collects the whole number of beasts that he intends to feed into one flock, and turns them all at once into one of these division; which, being quite fresh, and of sufficient length of bite, would please their palate so much as to induce them to eat of it greedily, and fill their bellies before they thought of roaming about, and thus destroying it with their feet. And if the number of beasts were so great as to consume the best part of the grass of one of these inclosures in one day, they might be allowed to remain there no longer; giving them a fresh park every morning, so as that same delicious repast might be again repeated. And if there were just so many parks as there required days to make the grass of these fields advance to a proper length after being eat bare down, the first field would be ready to receive them by the time they had gone over all the others; so that they might be thus carried round in a constant rotation.[2]

This very accurate description of short-duration high-stock-density grazing with variable recovery periods is an excellent reminder that much of what we know about grazing is not new. Rather, we have rediscovered venerable practices and are applying them in concert with new insights and new technology.

André Voisin's *Grass Productivity*, written in the 1950s, is another good reminder that much of what we know about grazing has been around for decades. Voisin used the term *rational* to describe good grazing systems. Rational thinking can be helpful in planning and managing grazing. Rational grazing can also refer to the importance of "rationing" out the pasture instead of releasing the herd or flock into the whole area. Voisin's book is full of technical information, but also shares his obvious passion for the subject, which he expresses here, along with a good description of what a well-managed pasture landscape should look like.

> What loveliness! What shades of colour all blending to form an even more magnificent picture where rational grazing is applied. The different paddocks, at different stages of regrowth, are not all of the same hue. Moreover, in a well-managed system the paddocks are not grazed in the same order as they stand, and so the colour tones, like reflections on the sea, do not gradually and uniformly diminish in intensity. Between two dark greens one glimpses a paddock lighter in colour, like the depth of a wave. A part where the grass has already begun to flower takes on an undulating, wavy aspect. What enchantment a pasture grazed in this way offers the eye![3]

The Result of Good Management

The result of good management is high-quality, dense, and highly productive pasture. What does such a pasture

Figure 5.1. Once pastures are being managed so that some are recently grazed, some are partially regrown, and others are fully regrown, there will be a variety of shades of green visible in the farm landscape. The most recently grazed pastures will be yellow or very pale green; fully regrown pastures will be darker green.

look like? It should include a diverse mix of several different grass species as well as several legume and forb species. This allows an intermingling of vertical grass leaves and stems with the more horizontal legumes and broad-leafed forbs. The dense growth makes it easier for livestock to get a full mouthful in each bite. It also creates more leaf area to capture sunlight, photosynthesize, and help the plant grow vigorously.

When you stand in a high-quality dense pasture and look down, you should see only plants with no bare soil visible between them. Imagine raindrops falling on this stand of plants. They will run down leaves and stems, making their way into the soil gently instead of crashing down on bare soil and causing compaction and erosion. Dense surface cover also provides an insulating

layer in winter and during conditions of extreme heat and dryness.

If you dig into a pasture that has high-quality soil, you'll find that it's full of deep and densely branching roots. There should be some litter on the soil surface, and under that layer of litter the soil will have an aggregate structure and plenty of organic matter. The soil should be full of biological activity, including both microorganisms and larger organisms such as earthworms and dung beetles.

With this image of a well-managed pasture as the goal, this chapter delves into management techniques that will prevent damage to pasture plants and instead ensure that they grow vigorously and create high-quality pasture.

Understanding Pasture Damage

Pasture damage is commonly caused by overgrazing, by rotating the livestock through the paddocks too fast (not enough regrowth time allowed before plants are grazed again), or by grazing plants down too short. Allowing winter access to pastures under the wrong soil conditions also damages pastures. Appendix A, with a table on troubleshooting pasture problems, can help guide you to an understanding of how some of the damage to pastures may have occurred.

OVERGRAZING DAMAGE

If a plant is grazed while it is still fueling growth from its stored energy reserves rather than from active photosynthesis, it is being overgrazed. Learning all the ways in which plants can be damaged by overgrazing is essential to managing pastures well. Once the problems caused by poor management are fully understood, they are easier to avoid. Overgrazing most often results from a few common mistakes, including:

- Taking down interior fences in the fall and letting livestock "clean up" the pastures when they are not fully regrown.
- Having a fixed rotational system of six or seven paddocks, with each grazed for one or two days, which doesn't allow for the regrowth period to be made longer when plant growth slows.
- Leaving animals in rapidly growing pastures for more than three days in a row, which allows them to regraze plants that are just starting to regrow.
- Returning animals to a pasture before all of the plants have fully regrown.
- Not adding extra acres of pasture to slow the speed of the grazing rotation when plant growth rates slow.

Plants in the early vegetative stage of regrowth, particularly if they have previously been grazed short, are very susceptible to damage if livestock graze them. Within just a few days of grazing, the healthiest and most vigorous pasture plants will begin to regrow leaves. However, unless some dark-green leaf material was still left on the plants, much of the energy for this regrowth

Figure 5.2. These plants were grazed three days ago and have already begun to regrow. If animals graze the new growth again at this critical stage, they can do severe overgrazing damage. The plants are not regrowing from the tips of the leaves, but are elongating leaves with regrowth from the collar at the base of each leaf.

will be drawn from stored energy in the plant. If animals continue to be allowed to graze the new growth, these plants will have to use even more stored energy to regrow a second time. Depending on the time of year and condition of the plant, this double whammy may lower the plant vigor and health for the rest of the grazing season, and possibly even lower next year's yield.

UNTOWARD ACCELERATION

In *Grass Productivity*, André Voisin discussed what happens when paddocks are not rested long enough between grazings. Voisin used the term *untoward acceleration* to describe what happens as each grazing of the paddock provides less forage than the last time. Because there is less forage, cattle are moved more quickly to the next paddock, and the regrowth period becomes shorter throughout the grazing season until most of the plants are overgrazed and there is little or no feed left.[4] The rotation gets faster when it should be going slower! Gregg Judy, a farmer from Missouri, has referred to this process as a "death spiral." An accurate description of what happens to many of the plants!

UNDERGRAZING?

Undergrazing is a term sometimes used to describe pastures that have been allowed to grow tall, and then are grazed only lightly with a low stocking density. This light grazing can be a useful thing to do on newly seeded pastures or pastures recovering from overgrazing damage or previous severe/short grazing. Undergrazing can leave a lot of ungrazed and nontrampled plant material, which shades the lower plants. This can create a problem, because sunlight does periodically need to reach the lower parts of the pasture to trigger the growing points of some of the grasses and legumes to produce new stolons and tillers. Continuous undergrazing of pastures that have grown fairly tall will result in lower plant density, and loss of lower-growing plants that can't tolerate so much shade.

WINTER AND MUD SEASON DAMAGE

Grazing in the winter by using stockpiled pasture can be a great way to reduce the need for stored forages. "Bale grazing," or feeding stored forages on pasture during the dormant season, can also be used to build soil fertility and employ hoof action strategically in a pasture. However, if soils are wet and not frozen when livestock have access to pastures, a significant amount of damage to plants and soils can result. Each farm will need to assess the local climate, soil types, and seasonal changes in soil conditions to decide what times of the year livestock should be kept out of some pasture areas.

Allowing livestock access to the pastures during the non-growing season can cause damage particularly during wet or freeze/thaw conditions. The longer the herd or flock is left in the area, the more likely they are to do damage. Soil compaction and damage to plant root

Figure 5.3. This pasture is becoming increasingly low in plant density. It has an accumulating layer of residue on the soil surface due to several years of growing too tall and being only lightly grazed with a low stock density.

Figure 5.4. This pasture shows damage that occurred to both the soil and plants during the winter when cattle trampled the area while the soil was wet and not frozen.

reserves will sometimes take the form of obviously visible "pugging" or deep trampling damage to soils and plants. However, sometimes it is less immediately obvious that soils are being compacted, but the result is slower spring growth rates and overall decreased pasture productivity.

In saturated soil conditions, grazing livestock may damage plant crowns and even roots through hoof impact and treading or pugging. This herd impact can also cause soil compaction.

Some pasture plant mixtures establish a denser sod that is better able to tolerate livestock treading during wet soil conditions. Once established, reed canary grass, for example, can create a sturdy sod that keeps livestock and equipment from sinking into the mud.

Depending on the local climate and seasonal precipitation, it may be important to plan ahead for "mud season" to avoid damaging pastures. Strategies can include setting aside some well-drained south-facing pastures for spring grazing in regions where there is a spring-thaw mud season. Lower-lying heavy soils can then be grazed later in the season once the sod is well established and soils are less saturated with water.

In some locations, it may be necessary to move livestock to a sacrifice pasture, heavy-use feeding area, feedlot, or barnyard and keep them entirely off pastures to prevent damage to soils and plants. In these situations, it's necessary to feed stored forages and to manage manure to prevent runoff and capture valuable nutrients so they can be later returned to cropland on the farm.

DROUGHT MANAGEMENT

Plants are not able to photosynthesize as much when they're drought-stressed or unable to grow new leaves. This makes them more dependent on their stored carbohydrates. During dry weather, check plants in your pastures regularly for signs of moisture stress, and adjust the grazing management. Leaving a taller post-grazing residual height means plants may not have to use as much stored energy to regrow new leaves. Heavy defoliation can also reduce root growth, which makes it harder for plants to compete for what little water and nutrients are available. Plants with more and deeper roots will do better in drought conditions. In addition, once the water is available again from rain or irrigation, it is very important to

allow drought-stressed or dormant plants plenty of time to regrow so they have time to refill their energy reserves.

As drought conditions continue, it is important to stay flexible and change the grazing rotation if needed to avoid damaging the pastures. It may be necessary to move cattle to a sacrifice pasture and feed them hay to prevent damage to pasture plants on the rest of the farm.

GRAZING PLANTS TOO SHORT

In addition to being damaged by being grazed while still regrowing, plants can also suffer if they're grazed down too short. Leaving livestock in a pasture so long that they graze plants all the way down to the base can allow them to eat the crown and growing points as well as depleting more of the plants' stored energy. Grazing too short also removes more of the green photosynthetically active leaves, so plants have to utilize more stored energy for regrowth and will regrow more slowly. This was discussed in chapter 3.

Figure 5.5. These horses are being strip grazed through the pasture to limit how long they are in each area, but horses are skilled at grazing the pasture plants very short. These plants will need some extra time to regrow due to this short post-grazing residual.

Smart Management to the Rescue!

With heightened awareness now of how poor grazing management can damage pastures, let's look again at the fundamental guidelines for good grazing management. These grazing guidelines include attention to the period of occupation and the length of the regrowth period after each grazing. Perennial grasses and legumes will respond to good grazing management with increased plant density, growth rates, vigor, and forage quality. This will create higher-quality pasture over time, while minimizing the expense of buying seed and tilling and reseeding the pastures. It may still be helpful to add some seed annually even to well-managed perennial pastures, but the cycle of having to renovate and reseed pastures due to declining quality won't be required.

Short Periods of Occupation

The first guideline to follow when you're planning and managing pasture is to plan for short periods of occupation — the amount of time the herd or flock is left in the paddock. A short period of occupation has several benefits for the plants. It reduces the risk that the livestock will graze the plants down too short and eat some of the crown or too much of the growing point of the plant. It also reduces the chance that the plants will start to regrow while livestock are still present, and thus minimizes their risk of having new tender leaves grazed off.

Moving the animals frequently is best, so that each paddock is not grazed for more than about three consecutive days. This "three-day rule" is not exact, but is a good place to start when managing cool-season humid climate pastures that are actively growing. The actual maximum length of the period of occupation will vary in different regions and with different plant species. But in general the faster the plants are growing, the shorter the period of occupation should be to prevent regrazing of actively growing plants. The most common way to shorten the period of occupation is to make the paddock smaller so that there is just the right amount of forage in the paddock for the planned period of occupation. In addition to keeping the plants happy, using smaller paddocks and moving the herd more often will also provide more consistent high-quality feed. Over

GRAZING GUIDELINES FOR GOOD MANAGEMENT

A well-planned and -managed grazing system requires close attention to the needs of the pasture plants and the livestock. Following these basic guidelines will help prioritize some of the most essential elements for success:

- Have a grazing plan!
- Allow plants enough time to fully regrow and recover after each grazing (variable recovery periods).
- Graze livestock in each area for a relatively short time to prevent regrazing (short period of occupation).

time this will help create higher pasture productivity and quality.

When animals are moved into a new pasture, the first bites they take will be the most palatable forage at the tops of the plants, which usually consists of the photosynthetically active leaves. This foliage is vegetative, highly digestible, and very palatable. Once they have consumed the top portion of plants in one spot, the animals will continue walking and biting tops off other plants. If they top and trample all the plants in a paddock but haven't yet filled their rumens, they will then walk back through the pasture, taking second bites as they go. When they take that second bite, they are more likely to be grazing the lower portions of the plants. If the paddock is too small or the animals are left in it for too long, they may even consume the crowns of some plants just above the soil surface.

Once a plant has had its leaves removed, it will allocate energy from storage in its lower parts up to the growing points. This is how it fuels regrowth of new leaves when much of the photosynthetically active leaf area has been grazed. This regrowth may happen rapidly when it is warm and there is good soil moisture, which might seem like a good thing. But not so if the grazing animals are still present! If the animals are still in the pasture as the plant regrows, they may return to the plant to graze it again, and in doing so may remove the new regrowth. The plant is then forced to deplete even more reserves (possibly all it has left) to try again to grow new leaves.

In this set of circumstances, the period of occupation needs to be short enough that none of the plants have time to grow new leaves and be regrazed by the livestock. Thus the faster the pastures are growing, the shorter the period of occupation should be. This is not intuitive! It might seem logical to think that when a pasture is growing rapidly, it would be better able to support a herd for a long period of time, because it is generating new growth fast. But from the plants' perspective, the goal is to protect that new regrowth when it first begins and is most vulnerable. In rapid growing conditions, some plants will regrow enough leaf material within just three days. Thus, an occupation period of three days may be all that pasture can handle.

During less favorable growing conditions, those same plants may take a week or longer to regrow enough leaf to be regrazed. But the three-day rule may still apply, not because of a need to remove the animals before new leaves can be regrazed, but to prevent them from grazing the plants down too short.

Remember that plants grazed short and severely will take longer to regrow since they have lost more of their photosynthetically active leaves. Severely grazed plants may also have had their crowns grazed off, including their growing points and the parts where energy is stored. Shorter periods of occupation mean that animals don't have time to graze plants too short, thus allowing pastures to improve more rapidly in both quality and quantity of plant growth.

VARIABLE RECOVERY PERIODS

The second essential principle is to vary the length of the recovery period based on how fast plants are growing. Variable recovery periods are essential to ensure that plants always have enough time to regrow. Monitoring how fast the pastures are growing is essential! As pasture growth slows, the speed of the grazing rotation must also slow down. During the grazing season, you must continuously assess how fast the plants are regrowing and adjust recovery periods. For example, regrowth period length may be as short as three weeks in early summer

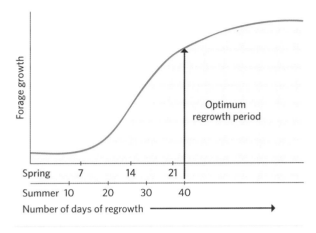

Figure 5.6. This graph illustrates how the amount of time needed for plants to fully regrow varies over the course of a growing season. Note that the numbers of days shown here are for example only. The actual number of days of regrowth needed will be different on each farm and in different regions.

when plants are growing rapidly, but it may be 40 to 60 days or longer later in the summer.

From the discussion of plant physiology in chapter 3, we know that plant height alone does not provide enough information on whether a pasture has recovered from grazing. We need to look more closely at what kinds of plants are in a pasture and what stage of growth they are in. Are the plants growing only vegetative leaves, or are the stems elongating? How many leaves are there? What color are they? Are there seed heads and stems forming?

We also need to understand what stage of carbohydrate balance the plant is in. Are there a lot of leaves? Is the plant still a light-yellow color, which indicates that it is mobilizing energy from the roots and crown of the plant and sending it to the plant tops? Or does that plant have lots of healthy dark-green photosynthetically active leaves—a sign that it is probably making and storing more energy than it is using?

As you've also learned, the amount of time it takes for a plant to regrow new leaves varies depending on how you have managed that plant in previous grazing cycles. Current temperatures and moisture conditions also play a role. It may take as little as 5 days for the plant to grow one new leaf, but it may take as long as 15 days or more in less ideal growing conditions. So if we want to wait for at least three or perhaps five new leaves to regrow it may take as little as 15 days or more than 60 days. Spring growth is usually faster than in midsummer, although the actual growth rates depend on the previous year's grazing management. If energy levels in the plants are low in the spring due to overgrazing the previous fall, spring regrowth can be very slow.

How can we apply these concepts in practical terms? Let's look at some farm examples. To provide the highest-quality feed to their cows, grass-based dairy farmers often give animals a fresh pasture after each milking. This provides the herd with a new area of high-quality feed twice a day. Providing highly digestible high-quality feed consistently is very important, particularly for livestock that are rapidly growing or producing milk. So this level of management is a priority for grass-based dairy farmers, as well as for farmers finishing lamb or beef animals on pasture.

Providing a new paddock twice a day also allows the pasture to be quickly grazed (short period of occupation!), after which the herd should be kept out of that paddock until it is fully regrown. This type of system requires either a lot of paddocks or a method of strip grazing. In addition, as the rate of regrowth slows, it is important to add acreage and paddocks to the system. For example, if pastures need 20 days to regrow in the spring, a system in which animals are moved twice daily will require at least 41 paddocks.

Figure 5.7. The beef herd at Harrier Fields Farm in New York State is being strip grazed through a high-quality perennial pasture. The front fence—a single strand of polywire supported by portable posts—is moved forward one or more times each day. A back fence behind this Devon herd is also moved every few days to prevent livestock from regrazing pasture they have already been in.

20 days x 2 paddocks/day = 40 paddocks
40 paddocks + 1 paddock (occupied on Day 1)
= 41 paddocks

Later in the summer these same paddocks will probably need 30 days or longer to regrow, so the system need will expand to 60 or more paddocks.

30 days x 2 paddocks/day = 60 paddocks
60 paddocks + 1 paddock (occupied on Day 1)
= 61 paddocks

Adding paddocks later in the summer is frequently done by adding more acreage to the rotation using land from which a first cut of hay was harvested. This requires timing the first cut of hay early enough to allow some

areas to grow back tall enough for grazing, and then using temporary fence to create new paddocks.

If setting up permanent paddocks isn't what the farmer wants to do, then strip grazing would provide the same good management. Strip grazing is done using portable posts and fencing such as polywire that are easy to move daily or even more than once a day, so that individual permanent paddocks are not needed. This system allows the same management control as the system of 40 to 60 paddocks described here, but with far fewer permanent fences.

If you don't want to either move fences frequently to strip graze or set up 40 to 60 fixed paddocks, don't worry! It is possible to use fewer, larger paddocks and move livestock less frequently. Beef or sheep farmers who farm part-time and also have an off-farm job will find that moving the herd or flock less frequently may work better for them. However, it is still important to avoid leaving

Figure 5.8. It can be helpful to pick and closely observe some individual grass plants as you assess a pasture. Count how many new leaves have grown on each tiller. The plant on the right has had time to grow more leaves and is closer to being ready to graze than the plant on the left.

the animals in an individual paddock for too long, and also to avoid returning the animals to regraze an area until it has fully recovered.

Let's look at an example of a system that uses fewer paddocks and longer grazing periods. A beef farmer who also has an off-farm job wants to move a herd only every three days. In the spring, when the regrowth period is about 20 days, the farmer would need seven or eight paddocks.

3 days/paddock x 7 paddocks = 21 days

But the farmer actually needs one more paddock since Paddock 7 needs to be resting while Paddock 1 is being grazed to allow the full 21 days of regrowth! So eight paddocks is the ideal number for this farm.

Later in the summer when regrowth slows to 30 days, an additional three to four paddocks could be added to ensure that plants have adequate recovery time.

3 days/paddock x 10 paddocks = 30 days

But remember the extra paddock, so the total needed is 11.

If you don't set up enough paddocks, but continue to rotate at the same speed, then this system will not work well for the plants or the animals! Pastures that are not as fully recovered will include more plants that have not yet grown enough leaves, and have not had enough time to fully recharge depleted energy reserves. So if cattle graze a pasture when it is in an earlier vegetative growth phase, the result will be more severe defoliation of plants, and more loss in annual forage production. This is why an extended period of occupation does more damage to a pasture that has had less time to regrow than to a pasture that has had an extra week or two of regrowth.

The bottom line is that regrowth periods must be variable! Chapter 3 has some additional information on variable recovery periods and strategies for managing the forage gap, which can happen at times of the year when pasture growth slows.

Pasture Improvement Strategies

Learning to apply the basic principles of good grazing management to your real-life pastures can be challenging, but you'll appreciate the differences in quality you see in your animals and pastures as your skills improve. But the basics aren't the whole story — you may want to learn additional strategies to improve your pastures even further. For example, what can you do if a pasture isn't growing high-quality forages over the full growing season as well as you want it to be? A common suggestion for "fixing" pastures is pasture renovation through tillage and reseeding. However, a full-scale pasture renovation is costly in both money and labor, so let's first look at lower-cost and lower-labor alternatives. Keep in mind, too, that no pasture improvement strategy can truly succeed if soil health is a limiting factor. If you haven't evaluated your soils (this was covered in chapter 4), it's smart to do so before you embark on any of the strategies that follow here. Appendix A on troubleshooting pasture problems may provide some ideas on how pasture problems began, and how to prevent and repair them.

FIX THE GRAZING SYSTEM FIRST!

Improving grazing management is the first way to try to improve pasture. So even if you think you've mastered the basics, review your management before you embark on a costly renovation. Also keep in mind that if you invest money and time to renovate and reseed a pasture, but your grazing management still isn't great, the pasture will rapidly return to the weedy nonproductive mess it was before!

Grazing management choices can cause a shift in the composition of plant species in a pasture. Under good grazing management, plants that are grazing-adapted will thrive, while poor-quality pasture plants (weeds) will die out. White clover will naturally spread by its stolons, grasses will spread by tillering, and woody weed species that do not like to be grazed frequently will grow less vigorously and die out.

Even overgrown brushy pastures can be improved in many situations without plowing and reseeding. To accomplish this, a large group of cattle or sheep (the mob) is used to rapidly defoliate pastures or brushy areas repeatedly to kill weed species and encourage pasture plant growth. This method of mob stocking or mob grazing requires a high stocking density and is best done with a group of animals such as nonlactating ewes or beef cows that do not have high-quality-feed needs.

FROST SEEDING

Frost seeding can be an effective strategy when a low-cost method of introducing new plant species is needed. This method of broadcasting seed onto the pasture is usually done in early spring when the soils are frozen at night and thawing during the day. The thawing and freezing action of the soil helps move the seed and soil into better contact with each other. Seed-to-soil contact is important because it keep seeds and baby plants moist and protected, which is favorable for germination and survival of small seedlings. Frost seeding works very well with legumes such as white and red clover. This can be done with just 2 to 3 pounds of seed per acre. These clovers can be frost seeded in late fall as well as in the spring.

Figure 5.9. These New York dairy cows are enjoying eating the leaves of a plant commonly considered to be a weed. Burdock plants, if grazed when the leaves are still tender and before the sticky "burrs" have developed, can be quite palatable.

Figure 5.10. The only burdock remaining in this pasture at Thistle Creek Farms in Pennsylvania is this bare stem. The beef herd has done an excellent job of defoliating all the burdock under high-stock-density grazing. Photo courtesy of George Lake, Thistle Creek Farms.

Grass is harder to establish by frost seeding than clovers. One reason for this is that most grasses have light seeds without the hard protective seed coating that legumes have. The lightweight grass seeds are more likely to get caught in foliage or plant residue instead of landing on the soil surface, so it can be harder to establish good seed-to-soil contact. In addition, the lack of a hard seed coating increases the chances the seeds will germinate at the wrong time (such as in the fall); freezing temperatures may then kill the tender young seedlings.

Try these tips to improve the success of frost seeding:

- Graze heavily in the fall to control grass growth and create more open soil; then sow grass seed onto the bare soil areas in the spring.
- Test soils and fix imbalances that hamper legume growth. Pay particular attention to lime, potassium, boron, and phosphorus.
- Seed in late winter (clovers only!) or early spring.
- Consider frost-seeding a mix of white clover and red clover, to add diversity to the pasture.

If you are including grass seed in the frost seeding strategy, spring seeding is required so new grass plants are not killed by freezing temperatures. Compared to the clovers, a lower percentage of the grass seed will successfully germinate and establish itself. Perennial ryegrass and orchard grass often establish better than other grasses. Even though the success rate may be low with the grasses, and the results may not show immediately, regularly adding some new seed to pastures can provide long-term improvements.

TRAMPLE SEEDING

Having livestock "trample" seed into a pasture may work in certain conditions. For this to succeed, soils need to be moist, and the trampling needs to be done quickly so that the seed is pushed into the soil, but then the livestock are moved out of the area before germination occurs. As with frost seeding, hard-seeded species (clovers) often show a higher success rate. Germination and survival of grasses will be variable with this method depending on temperatures, soil moisture, and how well the seed

Figure 5.11. This close-up view of a pasture that was trample seeded following a month of wet weather shows excellent germination of red clover, white clover, and some bird's-foot trefoil.

has made contact with the soil so it can stay moist and protected as it starts to grow.

I know of a farm that had a pasture where plant density was very low due to winterkill of many of the grass and legume species the preceding year. During a month of wet weather in the early summer, the farmers were able to spread a mix of legume and grass seed on the bare soil areas in that pasture before grazing. The beef herd trampled the seed into the soil and — due to the ideal soil moisture conditions — germination of new pasture plants was excellent, as shown in figure 5.11.

Neither frost seeding or trample seeding are an instant cure to a problem pasture. However, they are a useful part of an overall strategy to gradually improve the quality and productivity of a pasture.

No-Till Seeding

No-till seeding can reduce the risk of soil erosion when compared with full plowing and harrowing of a pasture. It may also allow more moisture to be retained in the soil, which can improve germination and seedling survival. No-till seeding may be done in areas where an annual crop was grown and harvested, or fields where a cover crop was previously grown and harvested, rolled, or winterkilled.

No-till can also be done in an established pasture sod, although the success rate is often lower. Not all species will establish well when no-till drilled into a pasture sod without significant control of the preexisting vegetation, particularly if the sod is dense. Control of weeds or other competing vegetation may require mowing, grazing, or other methods. In existing pastures, one way to control existing vegetation is by a severe fall grazing, followed by spring seeding. In some areas, using a high-stock-density group of animals to trample the sod can also help suppress the growth of existing plants that will compete with the new seedlings. Choosing the right variety, timing of planting, and seeding method will also increase chances of success. Timing needs to be managed so that seeds are sown when soil moisture and temperature conditions are favorable to increase seedling survival.

Adding new seed to an existing perennial pasture often does not produce immediate results. However, it may still be worth some regular investment in seeds of improved grass or legume varieties. Over time, some of these plants will gradually become established in the pasture and improve the overall productivity and quality.

Pasture Renovation

If you do decide to embark on a new pasture seeding, it is important to remind yourself (again!) that productive pastures are a result of good management, not just planting a new or improved species. Before writing that check to pay for tillage, fertilizer, and seed, make sure your grazing system is well planned and managed so that your investment isn't wasted. Tillage and reseeding take time and money, so be sure that you haven't overlooked the less expensive options for renovating and improving a field (discussed on page 68). Appendix C is a simple monitoring worksheet that can help you understand if the current grazing management is improving or actually damaging the pastures.

Once the decision to reseed a pasture is made, it is important to proceed with good attention to the details necessary to allow these new plants to thrive. Doing some research and planning will increase the likelihood that the investment of time and money in the new seeding will be worth it.

A successful reseeding requires the selection of the right plants, high-quality seed, good soil fertility, weed control for the new seeding, good seedbed preparation, and the right weather or irrigation conditions. The investment in a new seeding is not just the cost in money; it also includes labor and the time that the livestock must be kept off the pasture while tender new plants become established.

It may be necessary to begin planning a year or two ahead of time in order to reduce weed pressure in the field through tillage or other methods. Planning ahead allows time for soil testing and application of fertility inputs, too. The better the health and fertility of the soil, the better the chances of the new seeding establishing successfully.

Deciding What to Plant

Whether reseeding is necessary as part of a rotation with annual crops or for another reason, choosing what to

Figure 5.12. This photo shows how much bare soil there may be in a pasture after it has been completely tilled and reseeded. This combination of bare unprotected soil and young seedlings that don't yet have fully developed root systems requires careful management to prevent weed growth and erosion, and to encourage the desirable plants to grow vigorously.

plant is important. Selecting the right species, variety, or mix of species involves a consideration of soil pH, drainage, fertility, climate, crop palatability, the type of grazing livestock, weed pressure, harvest method, and the length of time the stand is needed.

The most critical consideration is to match the plants to the local climate and soils. What grows well a few hundred miles south of your farm may not grow well in your fields. Even minor changes in soil type or drainage from field to field can change what type of grasses and legumes will do best. Don't just plant what the local seed sales company recommends. Talk to the neighbors (the ones with good-quality pastures!) and take notice of what grows well year after year in their fields. Attend some local field days or pasture workshops to see which

forage varieties are doing well. As you narrow your list of choices, don't forget to ensure that the plants on your list do not include any that have toxicity or palatability issues.

A farmer reseeding a pasture in northern Vermont is likely to choose a different mix of pasture species than a farmer in Pennsylvania. For example, tall fescue might be planted on a Pennsylvania farm because it grows well in areas with hot dry summers, and can be used for fall/winter stockpiled grazing. However, in northern climates where it is cooler and moister in midsummer, it's relatively easy to grow grasses that are much more palatable than tall fescue. So tall fescue is not something the Vermont farmer is likely to choose as part of a summer pasture mix, but it is a good choice for farms with hot, dry summers or when used for stockpiled grazing.

Some pasture plants, such as reed canary grass, thrive in wet soils, while others will not survive such conditions. Reed canary, however, may be less palatable than other species of grasses in a pasture setting. Alsike clover can do well in wet soils, but it can be less productive than red clover and does not frost seed as easily. What is well suited for one region or farm isn't a good choice in an area with a different climate or soil conditions!

Many farms have land that is used for both grazing and haying, depending on the time of year. Conveniently, many perennial grasses and legumes do well under either good grazing management or hay management. However, if an area is going to be mostly hayed, then it's a good choice to plant taller-growing species such as timothy, bromegrass, red clover, or alfalfa, which are well suited to hay production. If the area will be primarily grazed, choosing plants that persist well under animal grazing and trampling — such as meadow fescue, Kentucky bluegrass, perennial ryegrass, and white clover — would be wise. See also chapters 3 and 6 for more discussion of how these plants grow.

Farmers who have access to equipment, labor, and suitable tillable land can rotate annual crops with a pasture grass/legume sod. Farms in areas with extended summer dry periods may grow summer annual crops such as sorghum sudan grass or millet for midsummer grazing. Farms may also benefit from grazing brassica crops or other alternative annual pasture mixes to extend the grazing season into the fall (you'll find more information about extending the grazing season at the end of this chapter).

When planting an annual crop for grazing, it may make sense to plant just one species such as millet or sorghum sudan grass. However, when planting a perennial pasture, it is always best to plant a diverse mixture of grasses and legumes. There may also be some advantage in including other forb plants in the perennial mixture.

A mixture will increase the chances that *something* will grow in all the different areas of a pasture: both wet and dry spots, as well as shaded and sunny areas. Diversity of species can increase the length of the productive grazing season, since some plants will grow better in the cool spring and fall weather, while others prefer hotter dry weather.

A mixture also helps reduce the chance of weeds in the pasture by reducing bare soil, which will more commonly occur in a monoculture planting. In addition, legumes provide nitrogen for the grasses. Different plant species have differently shaped root systems; some have deep taproots and others, shallower roots. Having a variety of depths and types of root systems helps maintain healthy soil and prevent erosion.

Chapter 6 includes details on the many types of pasture grasses, legumes, and forbs. Note that certified organic farms must comply with regulations on what types of seeds and seed inoculants (and fertilizers) are allowed. Refer to appendix D for more detail on this.

PREPARING THE SEEDBED

A correctly prepared seedbed allows good seed-to-soil contact. As explained on page 68, seed must touch the soil so that it can absorb moisture, which is essential for germination. This contact with moist soil also must continue after germination when the tiny new seedling is most susceptible to drying out. This is particularly important with the seeds of many of the grasses. In a poorly prepared seedbed, these light fluffy seeds may land on plant residue on top of the soil, where they will either be too dry to germinate, or may germinate and then be too dry to survive as seedlings.

When the pasture is fully tilled or renovated, it will probably be plowed, disked, and harrowed to control weeds and create a smooth, even surface for seeding. For pasture species that are difficult to establish, this may be the only way to ensure seeds will germinate and survive.

Soils should be worked enough to create a firm seedbed with sufficient loose soil on the surface to allow good seed-to-soil contact. Too much residue or excess large soil chunks can make it harder to get the seeds in close contact with moist soil. However, overworking the soil with too much tillage is bad for the soil structure, and creates seedbeds that will dry out quickly or form a surface crust.

A risk of this type of tillage and total pasture renovation is erosion from either wind or water. On sloped fields, tillage and planting should be done on a contour, and a companion or nurse crop should be used. Nurse crops such as oats or another annual species can used to

Table 5.1. Pasture Plant Seeding Rates

Species	Seeding Rate in Pounds of Seed per Acre (unless otherwise noted)	
	Planted Alone	Planted in a Mix
Kentucky bluegrass	12–14	4–6
Orchard grass	8–12	3–5
Perennial ryegrass	10–20	5–6
Reed canary grass	8–14	4–8
Smooth bromegrass	12–18	6–8
Tall fescue	12–14	6–12
Timothy	8–10	2–8
Big bluestem	10–12	
Indian grass	10–12	
Eastern gama grass	8–10	8–10
Switchgrass	8–10	
Alfalfa	12–20	2–10
Alsike clover		2–5
Bird's-foot trefoil	7–10	2–8
Red clover	2–6 (frost seeded into existing pasture)	2–6
White clover	1–4 (frost seeded into existing pasture)	1–4
Peas with oats	Peas, 90–100; oats, 50	
Annual ryegrass	20–25	10–15
Millet		15–40
Sorghum sudan grass		25–40
Spring small grains	3 bu/acre	1.5 bu/acre as a nurse crop
Brassicas	Turnips 1.5–2; kale or rape, 3–4 (drilled into existing pasture)	turnips 1.5–2; kale or rape, 3–4
Forage chicory	3–4	1–2

help protect a new seeding and the soil. In addition, leaving some residue from a previous crop on the surface will also reduce the risk of erosion. Some fields may have such a high risk of erosion that complete tillage for reseeding is not a good choice, so frost seeding or no-till may be a better option.

SEEDING SUCCESSFULLY

Once the seedbed is prepared, then seeding can be done. Timing of planting is critical, so that seeds go into the soil when soil moisture and temperature conditions are optimal for germination and seedling survival. Each season and region present their own unique challenges. In some areas, planting too early in spring when soils are too cold may result in seeds rotting instead of germinating. But if you wait until too late into spring or summer, soil surfaces may dry out quickly after seeding, which can stress and kill the young seedlings. Fall seeding is an option in some regions and for some species. However, this may not be successful if the seedlings are not well established when soil temperatures drop or the first frost occurs.

Clearly the timing of seeding varies depending on the local climate and soil conditions on the farm. A good resource to learn about local recommended planting dates in your area is your local extension or NRCS office. These organizations may have fact sheets or guidelines for local planting dates, rates, and even well-adapted local varieties.

The seeding itself may be done with a drill or by broadcasting. Using a drill will usually improve the chances of good seed-to-soil contact and germination compared with broadcast seeding. The addition of a cultipacker or roller can further improve seed-to-soil contact.

When reseeding a pasture, the ideal is to seed at a rate that results in 60 to 80 new plant seedlings per square foot. However, the amount of seed used per acre, and the depth it should be planted, will vary due to several factors.

Seed planting depth varies according to the plant type and the size of the seed. Larger seed generally needs to go deeper, while tiny seed will struggle to emerge if it is too deep in the soil. Refer to the specific information from the seed sales company about the plant or pasture mix. The ideal depth and rate of seeding depend on the seeding method in use as well.

Soil texture plays a role here, too, because seeds can emerge more easily in light soils than heavy ones. When seeding into lighter soils, it may be possible to use a lower seed rate than you might have to use in a heavier soil.

Seeding at a lower rate may be more successful in well-prepared seedbeds than in more roughly prepared soils. Seed quality, as well as the expected germination rate for a particular species or variety, also helps determine seeding rate. If seeds have lower-than-usual viability, a higher seeding rate is needed. When seeding a mixture of many pasture species, or adding in a nurse crop species, some adjustment in pounds of seed per acre will also be needed.[5] Refer to table 5.1, information from your seed supplier, or appendix E for resources which supply more detail on setting seeding rates.

SPECIES-SPECIFIC PLANTING DETAILS

As mentioned earlier, grasses and legumes have different types of seeds, and also have different growth habits. In addition, warm-season plants, cool-season plants, annuals, and perennials grow differently, so each of these plants will do best when planted with the timing and method best suited to its individual needs.

Grasses

Cool-season grasses are usually planted in the spring or summer when soil moisture and temperatures are ideal. Warm-season grasses, which are slower to establish, won't compete well with weeds and other grasses. Because of this, it's best to plant warm-season grasses where there is low weed pressure, and then manage with extra care for the first two years, which is how long most stands take to establish. Depending on the region and climate, either spring seeding or a fall planting of dormant seed is an option for planting warm-season grasses.

Consider seed size when deciding on seeding depth. Grasses such as eastern gama grass, which have a larger seed, should be planted ½ inch deep. Smaller grass seeds, such as bluegrass, should be seeded only ¼ inch deep to allow successful emergence from the soil.

Small Grains

Small grains grown as an annual crop for grazing are generally planted at a higher seeding rate than when planted for a grain crop. A crop for grazing can also be planted earlier than usual for the species, but soil temperatures need to be high enough to allow germination. There may be some benefit to planting a mixture of several species of small grains or adding a legume to increase diversity and help extend the grazing period.

Legumes

Legumes vary widely in seed type and size, and each can have very different seeding requirements. Large seed forages such as field peas and other annual legumes need to be planted deeper than small hard-coated seeds such as white and red clover.

Legumes don't require nitrogen fertilizer at planting, but will benefit from being inoculated with the correct *Rhizobium* species. If the seed is not pre-inoculated, purchase a fresh inoculant at seeding time. Be sure to select the right type of inoculant based on the legume species being planted. In addition, if the farm is certified organic, make sure the inoculant (as well as the seed) is an approved non-GMO product. Chapters 3 and 5 include more information on these helpful nitrogen-fixing organisms. Appendix D has more information on organic certification.

Figure 5.13. Different plants have a range of seed sizes, which must be taken into account when deciding how deep to plant them. Here two legumes show the range from large pea seeds to the very small clover seed.

Extending the Grazing Season

As you work on improving pasture quality, it is also smart to consider how you can start extending the grazing season. Anyone who is not convinced of the benefits of season extension should read *Kick the Hay Habit* by Jim Gerrish.[6] Jim discusses the benefits and methods of grazing longer, or even grazing year-round in a wide variety of climates. Year-round grazing is obviously not possible where severe winter weather and a deep snowpack are the norm. And as discussed earlier in this chapter, care must be taken not to damage soils when they are saturated. However, the longer the grazing season, the less stored forage you'll need to make and feed.

Harvesting, storing, and feeding out forage is a lot more expensive than grazing. Ed Rayburn wrote that "more than half the cost of producing livestock is feed cost, mostly winter feed ... Extending the grazing season is one way to reduce winter feed costs; pasture costs one-third to one-half as much to produce as harvested feeds.

By striving to achieve as close to a 12 month grazing season as practical, feed costs can be minimized."[7]

Forage costs can include purchase and delivery costs or the costs of harvest, fuel, and equipment maintenance as well as the labor and expense of moving, storing, handling, and spreading manure. Figuring out every possible way to keep the livestock out on pasture harvesting their own feed and spreading their own manure is an obvious way to decrease farm labor and increase farm profits.

MANAGEMENT CONSIDERATIONS

Farms in some mild climates may be able to graze year-round. In areas with icy, very cold, or snowy winters, however, there will be limits on how long the grazing season can be extended. For example, stockpiling forage is an important part of extending the grazing season, but there may be a loss of stockpiled forage under the snow or ice. There is also the risk of loss of animal productivity or animal welfare concerns due to grazing in some challenging weather conditions. With high-performance animals such as lactating dairy cows, it may be impractical to extend the grazing season too late in snowy or icy conditions.

Advanced planning is important for many aspects of season extension. For example, farmers may want to plan ahead to set fence posts into the ground before the soil freezes. Alternatively, you can use tumble wheels to move fence over frozen ground.

Allowing grazing animals to tread on saturated soils or soils that are freezing and thawing can result in significant damage to perennial plant crowns, plant roots, and soil structure, as shown in figure 5.4 earlier in this chapter. Pastures with a dense, well-developed sod will be better able to withstand grazing during wet conditions. Once the soil is solidly frozen, grazing can be done with less risk of compaction or damage to plant crowns. Livestock can graze through a few inches of snow, as long as it is not too deep, does not have an ice crust, and is not too compacted.

Extending the grazing season can be done with the existing perennial pasture plants on the farm, or it can be done by adding annuals. With either annuals or perennials, careful planning is also required to ensure there will be enough of the right quality and quantity of feed available throughout the extended grazing season. Because of the additional cost and labor of planting annuals, using more perennials is always a more cost-effective way to extend the grazing season. However, avoiding overgrazing damage to perennial plants by grazing them too often or at the wrong time is crucial.

As explained in chapter 1, one of the benefits of improved grazing management is healthier plants that not only produce more high-quality forage but also produce over a longer growing season. Perennial pasture plants that have been well managed will grow more vigorously in the spring, are more likely to persist through the midsummer heat, and will grow later into the fall.

By carefully planning the grazing so that every pasture on the farm gets appropriate regrowth periods and plants don't get grazed down too short (particularly in the fall), the grazing season will naturally start to become longer. Farms that typically started to run out of pasture in mid- to late summer when they had a poor grazing system will find that they don't have to start supplementing the livestock with hay as early each year. Over time,

this can add several weeks or even months of pasture to the annual forage supply.

Stockpiling Perennial Forage

Stockpiling is the practice of allowing pastures to grow and accumulate forage for use later in the year. For cool-season perennial pastures, this is usually done by taking the last harvest or grazing at least 50 to 75 days before the end of the growing season. This allows the late-season forage—which in cool-season grasses usually consists of mostly high-quality leafy grass and legumes—to accumulate. The plants have plenty of time to photosynthesize and store energy.

While any cool-season perennial pasture plant can be stockpiled for fall grazing, some serve the purpose better than others. Tall fescue and bird's-foot trefoil are well suited because they retain their leaves and will continue to grow in cool fall temperatures. Bird's-foot trefoil is also a non-bloat-causing legume. If you're planting tall fescue for fall grazing, be sure to use the more palatable endophyte-free varieties. (See the Fescues section of chapter 6 for an explanation of endophyte-free varieties.)

Other grass and legume species are more likely to lodge (fall over) or lose their leaves after a frost. For example, perennial ryegrass, orchard grass, alfalfa, and red clover can be stockpiled, but they may lose more leaves and decrease more in quality than fescue and bird's-foot trefoil. In addition, care must be taken when grazing bloating legumes such as red clover in the fall after frosts.

Fall stockpiling requires planning ahead to end the grazing or harvest of a pasture so that it can start to regrow at the time. This timing may be when pasture is in short supply on the farm, which is why planning ahead is needed to assure that the stockpiling happens correctly. A full season-long grazing plan will be particularly helpful in this process. This allows all the fields and the grazing groups and haying activities to be planned out so the fields to be stockpiled are assured the long growth period needed.

For many farms in milder climates in the United States, well-planned grazing with perennials can provide most or all of the forage for the livestock on the farm. However, depending on the type of livestock, their

Figure 5.14. Bird's-foot trefoil, growing here in a mix with cool-season perennial grasses and some red clover, is well suited for fall stockpiling.

Figure 5.15. The dairy herd at the Beidler farm in Vermont begins grazing a strip of millet pasture. The millet is grown as a midsummer annual that can be grazed or harvested as a forage or seed crop.

nutritional needs, and the local climate, making and feeding stored forages may also be necessary.

Annuals for Grazing

Another strategy to extend the grazing season and reduce the need for stored forages is growing annuals. This may include winter grains planted in the fall and grazed in the spring in areas with cold snowy winters. It may also include grazing these annuals during the winter where the climate makes this possible. Fall grazing of annuals is possible in cool northern areas as long as they are not covered with too much snow and ice.

In areas where there is a mid grazing season forage shortage when cool-season perennial pasture plants go dormant or grow too slowly, other species such as warm-season grasses or annuals can be grazed. Warm-season plants such as millet or sorghum sudan grass can provide a lot of high-quality dry matter per acre during the time of year when cool-season perennial pasture growth slows. Other mid-grazing-season annuals for grazing include oats, or oats with peas. Grazing these annual crops requires planning ahead to do the weed control, create the right soil fertility, prepare a seedbed, plant, and give the plants the right amount of time to grow.

If grazed correctly, most of these annual summer grazing crops should be able to provide multiple grazings. It is important to select forage varieties of these plants, then graze them carefully so the plants can regrow. Millet does not pose a risk of prussic acid poisoning, but care must be taken with the sorghum sudan grasses. See chapter 12, Health Concerns, for more on this topic.

Annual forages that grow well in the fall include small grains such as oats, and forage brassicas. Winter grains such as wheat, barley, triticale, and rye can be planted as a monoculture or in a mix of several types. Fall brassicas include turnip, kale, and rape. These are usually planted in a mix with other plants, or they can be drilled into an existing pasture.

Test soil fertility before planting any of these annual crops. These test results may tell you that manure or other fertilizers are needed to assure a successful crop. It is also important to time planting so there is enough soil moisture when seeding, and there is enough time for forage to accumulate by the end of the growing season.

Forage brassicas take 60 to 80 days from planting before they are ready to graze, so the timing of planting and grazing is important. It's also important to feed them carefully — and note that they should not be grazed as

Figure 5.16. The herd at the Beidler Farm has quickly consumed much of the millet in the strip they were moved into less than an hour earlier. This paddock will regrow and be ready to graze again in one to two weeks.

the only forage! These should be grazed for only a short time each day, and animals should have access to stored forages or grass/legume pastures if grazing a brassica monoculture. Ideally they should be grazed as part of a mixture. See chapter 12 for more detail on issues from too much brassica in the ration.

Annuals or Perennials in Spring?

Some cereal grain crops can be overwintered and grazed in the spring. These winter grains include winter rye, triticale, and wheat. Winter rye is generally the first one to break winter dormancy and begin to grow. Spring grazing on annual crops results in soil damage if the spring soil conditions are wet due to the risk of soil compaction. So if the field is likely to be wet in early spring, it is not a good candidate for grazing early annuals.

Many farms report that using annuals for spring grazing is less effective than other strategies to increase spring growth in perennial pastures. Perennial pastures have a better-developed sod so there are fewer issues with wet spring soils. In addition, the cost of well-managed perennials is significantly lower than that of annual crops!

To encourage more spring growth in perennial pastures, plants must be given long recovery periods during the previous grazing season. In addition, you must carefully protect plant crowns and all the new tillers in the perennial grasses by making sure plants are not grazed too short in the fall.

Strip Grazing Stockpiled Pasture

Free access by the livestock to fall stockpiled pasture will result in trampled and wasted forage. If this occurs during wet weather, waste and treading damage to the plants can be high. A better method is to give them just what they will graze in a day or less and move them more often.

Strip grazing fall stockpiled annuals or perennials is an ideal way to efficiently graze the crop. This also makes it easier to limit intake of plants such as brassicas to make sure the animals are getting a well-balanced ration. Strip grazing prevents the animals from trampling forage, which will then be wasted. Careful strip grazing in some cases may make it possible for the fall crop to be grazed more than once. Grazing rape or turnips quickly and lightly (leaving plenty of residual) can allow them to regrow for a second grazing.

The Art of Good Grazing

Grazing (Almost) Year-Round in the North

George Lake is the owner and manager of Thistle Creek Farms in west-central Pennsylvania. The farm was established in the 1920s, and under George's management it has been producing and direct marketing high-quality dry-aged grassfed beef for over 20 years. During that time, George has been working to steadily increase the length of the grazing season and the carrying capacity of his farm by improving soil health, forage production, and cattle genetics.

With a goal of grazing year-round, or as close to that as weather conditions will allow, the farm has invested in a very diverse mix of both annual and perennial forages while also building soil health and fertility. The beef herd at Thistle Creek Farm is rotated into new pasture at least once a day to keep the finishing animals gaining weight steadily. This grazing management is also building healthier soil and improving pasture productivity and quality.

During some winter conditions, feeding stored forages is necessary, so some forages are harvested from both the annual and perennial crops on the farm. Winter conditions that make grazing challenging for the cattle include ice crusts and deep snow, which prevent the herd from being able to harvest enough pasture dry matter. During these conditions, hay or balage is fed out to the herd.

With the animals outside year-round, the pasture water system has been designed to work even in freezing weather. Water lines are buried, and water tubs are engineered with overflows so they don't freeze.

Annual crops grown on the farm include forage sorghum, corn, and sorghum sudan grass. Perennial forage crops include a diverse mix of grasses and legumes. Plant species grown are all non-GMO varieties and are selected based on the soil type and growing conditions in the different fields. Some of the higher, drier areas include alfalfa in the mix. In the lower fields where soils are wetter, clovers are planted with meadow fescue. This diversity of plant species on different parts of the farm assures that something is growing during the entire grazing season. The cattle turn the high-quality forage into

Figure 5.17. In the last few years, George Lake has added more freeze-proof water tubs to the farm to make year-round grazing and out-wintering easier.

Figure 5.18. A highly productive, high-density mix of several varieties of meadow fescue, perennial ryegrass, Kentucky bluegrass, timothy, and white clovers grow in the pastures with wetter soils at Thistle Creek Farm.

Figure 5.19. Even in deep snow, the corn stays upright so it can be strip grazed by the mother cow herd. George uses his well-trained cattle dogs to hold the herd back while he sets up the next fence so he can give the herd a new area to graze. Photo courtesy of George Lake and Thistle Creek Farms.

high-quality beef that chefs are eager to buy to serve in regional restaurants.

Thistle Creek Farm has made a commitment to grass-finishing the beef cattle without grain. The finishing animals are fed only vegetative forages, so once ears form on the corn crop, only the mother cow group grazes on it. The standing corn has provided an excellent high-energy feed source for cows to graze in the winter when other standing forages are too deeply buried in snow.

In addition to over 400 head of beef cattle, the farm also has a flock of about 120 ewes. George is using the flock to improve some of the marginal areas of the farm, using grazing to clear brushy areas. The sheep are producing a healthy lamb crop while defoliating small trees, weeds, and brush. He plans to continue to grow the sheep flock as the land productivity in those areas continues to improve.

For both the beef and sheep, grazing is done with a medium to high stock density using permanent perimeter fences and interior portable fencing. With the assistance of his well-trained working dogs and the portable interior fencing, George moves the animals forward into fresh pasture from two to five times a day.

The back fence is moved forward less often than the front. This creates a period of occupation in each area of three to five days so the animals can walk back to the water tub. In recent years, George has been adding more water tubs so that the period of occupation in each pasture area can be shorter. This will allow the pasture plants to begin their rest and regrowth faster after each grazing, which will further improve pasture productivity.

In addition to fields of annuals or perennials for grazing or stored forage production, there are also some shade and shelter pasture areas. George established these

Figure 5.20. Moving the front fence forward several times a day allows this group of finishing animals access to a steady ration of high-quality pasture. The animals do a great job of trampling and spreading manure evenly, which helps continue the pasture improvement cycle.

areas many years ago. The trees are spaced far enough apart to allow sufficient light penetration to support some forage growth. The trees provide a cool shady area for cattle in hot weather, or shelter on cold windy days. Figure 4.9 shows one of the shade pastures on this farm.

George says that the most important tool for improving his pastures is the herd's hooves. When managed with the correct stock density, and applied to pastures that have had a long rest and regrowth period, the herd's hooves quickly trample plant residue and manure into the soil. This is indeed a powerful tool for improving plant productivity and soil health. The good results of long recovery periods, short periods of occupation, and well-managed trampling with hundreds of hooves show clearly at Thistle Creek Farms in both the high-quality pasture and highly sought-after grass-finished beef.

Common Pasture Plants

Pasture plants include a range of species including grasses, legumes, and the broad-leafed plants known as forbs. Some species do better in one region or climate than another. Before you buy seed for seeding or reseeding a pasture, make sure that the species is well adapted to the local climate on your farm. The information in this chapter on specific pasture plants will be helpful, but it's also important to do some additional local research to ensure that any new plant species you add to your pastures will grow well and will be palatable to the livestock.

Some pasture plants are generally considered weeds, but a weed on one farm may be considered a valuable forage species on another. Some "weedy" species can actually provide good-quality pasture when managed well as part of a diverse pasture mixture. For example, quack grass is a weed in the vegetable garden, but in a pasture can be a useful cool-season forage species. Other grass species, including annuals like crabgrass, are not as productive and high-quality as many of the ideal pasture species, but they may still provide some part of the overall pasture diversity.

In this chapter, I've summarized information on more than 40 common pasture plants. I've included a wide range of the common "improved pasture forage species" that can grow throughout much of North America, but it is by no means a comprehensive listing of all the plants you will encounter! I have grouped plants by category — grasses, legumes, and forbs, with even the so-called weeds included in their botanical category. Some of these plants are native to North America, but many were introduced from Europe and have naturalized in much of the United States and Canada, particularly east of the Mississippi River.

The information here is not intended as a plant ID guide or an in-depth discussion of plant care, but rather as a resource to help you learn more about many of the most common pasture plants. I've included information on what conditions they are best suited for, and the pros and cons of including them in a diverse pasture mix.

I consulted a wide range of references to compile the information in this chapter, many of which are listed in appendix E, which also includes other helpful books on pasture plants and online resources on pasture plant species and identification.

The Grasses

The grasses include both warm- and cool-season species. In addition, it may be surprising to find that some grasses that reappear year after year in a pasture are actually annuals that must regrow from seed each year. This section will first cover the cool-season perennials, then the warm-season ones, followed by the annuals.

COOL-SEASON PERENNIAL GRASSES

As discussed in chapters 3 and 5, cool-season grasses are most productive during the cooler, moister weather of spring and early summer. These plants produce a large volume of forage dry matter as they form seeds, after which they produce a lower volume of dry matter that is more vegetative (leaves). Some species may slow their growth or go dormant during very hot, dry midsummer conditions.

Bromegrass

Bromegrass (*Bromus catharticus*), also called smooth bromegrass, is a tall, sod-forming, high-yielding grass. It can be used in pasture or as a hay crop. It is one of the grass species that elevates some of its growing points, as explained in chapter 3, so it will not persist well in pastures under close or frequent grazing. It has a relatively high nutritive value and is palatable.

Figure 6.1. Bromegrass thrives at Earthwise Farm and Forest in central Vermont, where the farm owners plan for a tall pregrazing height, leave a tall residual, and provide enough time for the grass to fully regrow before it is grazed again.

Figure 6.2. Tall fescue's bunchgrass growth habit stands out among clover and other nonbunching grasses.

Brome spreads underground by rhizomes and will develop a deeply penetrating root system that, when managed well, will fill the soil with a mass of roots and rhizomes. It is suited to deep loamy soils, silt, or clay but will also do well in light, sandy, well-drained soils. It does not do well in alkaline or saline soils. It is highly tolerant of cold temperatures and can survive some temperature extremes and moisture stress. It mixes well with red clover or alfalfa in pasture and hay fields.

Because of its growth habit, this grass prefers a taller pregrazing height (at least 10 inches). It also does best with long regrowth periods between grazings.

Fescues

The fescue (*Festuca* spp.) genus includes several species that are quite different from one another. Tall fescue can provide fall or winter stockpiled pasture and grows well through hot, dry weather, but it may present some toxicity and palatability issues. Meadow fescue doesn't have these problems, and is well adapted to wet soils and cool, moist areas. There are also several fine-leafed fescues that are able to tolerate severe grazing but are not high yielding.

Tall fescue (*Festuca arundinacea*) is a deep-rooted grass that has a bunch-type growth habit but makes a dense sod. It has a large root system, and is one of the most drought-tolerant cool-season perennial grasses.

It can tolerate seasonally wet soils and survives soils that are acidic or alkaline. Due to its tolerance of a wide range of soil conditions, it is frequently used as a conservation planting to revegetate disturbed areas on roadsides and around mines.

There can be palatability issues with this species as well as health concerns when grazing varieties infected with a fungal endophyte. The endophyte is a microorganism that lives inside the plant and makes a toxic alkaloid. Many of the older varieties of tall fescue such as Kentucky 31 and those found in many conservation seed mixes are infected. However, there are endophyte-free varieties of tall fescue, and its ability to grow during hot, dry weather makes it a good choice for farms that are unable to grow the more palatable cool-season grass species during midsummer. Tall fescue can also provide good-quality fall stockpiled pasture for grazing because it will continue to grow later into the fall and will not lose its leaves as rapidly after frost compared with other cool-season grasses.

Because tall fescue is less palatable than other cool-season perennial grasses, it can become problematic and even somewhat invasive in a cool, moist region when planted in a mix with other cool-season species in a pasture system that allows for selective grazing. Over time the cattle will graze the other species, leaving the fescue to go to seed and spread through the pasture. For this reason, cool-season grasses other than fescues are a better choice for a diverse pasture for midsummer grazing in regions where drought and heat are not common.

The potential health problems or fescue toxicity that result from grazing endophyte-infected fescues include fescue foot and poor animal performance, which may be referred to as summer syndrome. Fescue foot symptoms include lameness; dry gangrene in the tail, legs, and ears; and a rough hair coat. Severe symptoms are rarely reported to be a problem except when grazing pure stands of tall fescue. The poor animal performance symptoms include lower rate of gain, a rise in body temperature, and poor performance during the hotter summer months.

However, in the big picture of the pasture ecosystem, the endophytes are not all bad! They are actually part of what protects tall fescue from heat and drought. Still, for the well-being of livestock, it's best to select one of the newer endophyte-free varieties that have higher palatability.

Fine-leafed fescues, which include sheep fescue (*Festuca ovina*) and red fescue (*F. rubra*), are most likely to be found in pasture seed mixtures for horse or sheep pasture, as they are able to tolerate severe grazing better than many other grasses.

The fine-leafed fescues do not contain endophytes (see the tall fescue description on page 83). However, they are not highly productive pasture plants, so they're more likely to be found in conservation mixes or seed mixtures intended to survive severe grazing and trampling.

Meadow fescue (*Festuca pratensis*) is smaller than tall fescue, is broad-leafed, is well adapted for grazing, and does not have the palatability or endophyte toxicity issues of tall fescue. Meadow fescue is a hardy bunchgrass and can do well in wet soils, though it will not tolerate extended flooding the way reed canary grass can.

It is well adapted to northern cool climates. Unlike tall fescue, it will not continue to grow in hot, dry midsummer conditions. Meadow fescue is also lower yielding than tall fescue.

This grass used to be more common, but went out of fashion (so to speak) and was largely forgotten by the mid-1950s. This may have been due to its susceptibility to some diseases. Recent research and variety selection work has resulted in some new varieties and growing interest. It is now becoming quite popular in seed mixes for wet pastures.[1]

Festulolium is a hybrid cross of fescue and ryegrass. There are several varieties available, and they exhibit significant differences from one another. This cool-season perennial is quite palatable, but not all varieties are winter-hardy. Growth habits also differ, so before including it in a pasture mix, do some research to determine which variety will be the best choice.

Kentucky Bluegrass

Kentucky bluegrass (*Poa pratensis* L.) is a low-growing, sod-forming grass found in pastures throughout the United States. It is so common in pastures that some people refer to it as a native plant, but it's actually from Europe, where it is commonly called smooth meadow grass.

It's well adapted to grazing because it is low growing and spreads by rhizomes. It creates a dense sod and can tolerate short grazing. It is very palatable and grows well with white clover. It is very winter-hardy and grows well in the spring and fall. However, it is not tolerant of high summer temperatures or drought, so summer growth can be limited in some areas.

When grown with white clover, this grass creates a dense pasture that is shorter than many of the other perennial grass species. It can be grazed at shorter pre-grazing heights than many of the other grasses.

Orchard Grass

Orchard grass (*Dactylis glomerata*) is known as cocksfoot in Europe. This tall-growing bunchgrass does well under managed grazing or haying. It is more drought-tolerant than Kentucky bluegrass and timothy, but not as tolerant as tall fescue. It has reasonable winter hardiness, but in northern areas cold winters without snow cover may limit growth. Orchard grass can persist in less fertile soils than timothy or Kentucky bluegrass can. It can also tolerate moderately poor drainage, but won't survive flooding or wet soils as well as some other grasses.

Because this plant stores energy in the bases of the lower stems, grazing it severely or short can damage plants. If plants are damaged from being grazed too severely, stands can thin out over time, leaving bunches of grass with bare soil or other possibly weedy species between.

It has a fibrous root system that is not as well distributed as some other pasture grasses such as bromegrass. The roots are generally deeper than Kentucky bluegrass or timothy roots, which helps with its drought tolerance.

Many orchard grass varieties mature early in the spring. After the seed heads have developed and been removed by harvest or haying, however, growth for the rest of the year is all high-quality leaves. It grows well with red clover or alfalfa, and later-maturing orchard grass varieties are often easier to manage in pastures or mixed alfalfa/grass fields. This can be an excellent pasture grass when well managed, but can have reduced palatability if allowed to become overmature.

Figure 6.3. Orchard grass is a tall-growing bunchgrass, shown here amid a stand of white clover.

Perennial Ryegrass

Perennial ryegrass (*Lolium perenne* L.) is a high-yielding, nutritious bunchgrass. It is a high-energy forage that does well under grazing in cool, moist regions. It grows best on fertile and well-drained soils and will not tolerate alkaline or very infertile soils.

It is not as winter-hardy as other cool-season forage species and it does not tolerate drought well. Due to the marginal cold hardiness of some varieties in more northern areas, it works best in a mix with other grass and clover species.

Diploid (double chromosome) and tetraploid (quadrupled chromosome) types of this grass are available. If you're looking for a variety with more cold hardiness and longer persistence, the diploid types may be a better choice. However, if a shorter-lived stand will work well, or if your farm is located in an area with milder winters, the tetraploid varieties have higher digestibility, tend to grow better when interseeded with legumes, and have larger leaves.

Some farmers have had luck frost seeding perennial ryegrass into pastures. Like all the grasses, however, it does not frost seed as well as legumes do.

Quack Grass

Quack grass (*Elymus repens* L.) is a sod-forming cool-season grass that spreads by long-lived white rhizomes. Due to its extensive growth of rhizomes and thick sod development, it is excellent at holding soil. Gardeners will recognize it as an annoying fast-spreading weed; it's considered a noxious weed in some regions. The seed heads can be confused with ryegrass seed heads, and seeds may contaminate cultivated forage seeds.

In a pasture or hay field, quack grass can be a useful part of the diverse mix of forage plants. It can grow in a wide variety of soil conditions, can withstand salinity and alkaline soils, and is fairly drought tolerant. You won't see quack grass listed on the label of your pasture seed mix, though it is very likely to appear naturally in your pastures and if managed well can be a nice addition to pasture diversity.

Reed Canary Grass

Reed canary grass (*Phalaris arundinacea* L.) is a tall, sod-forming, coarse, high-yielding grass. It is known for its tolerance of flooded, wet, and poorly drained soils, but it is also fairly drought tolerant. This species can grow even during the summer slump when other cool-season grass species are less productive.

This grass may be slower to establish than some other grass species. Once established, it spreads underground by rhizomes, and creates a dense, heavy sod in well-managed stands. The sod can be so firm that it can support grazing animals and farm equipment even in very wet conditions.

There can be issues with palatability of this grass, particularly if it is allowed to grow too tall. Some varieties have low palatability due to presence of alkaloids, but there are lower-alkaloid varieties available. Because reed canary can grow quite rapidly, it can be challenging to manage this grass in a diverse pasture mix with other grass species that require longer regrowth periods. By the time those species are ready to graze, the reed canary grass may be quite tall and have low palatability.

Figure 6.4. When reed canary grass is allowed to mature fully, as shown here, it can become less digestible and palatable.

Like smooth bromegrass and timothy, reed canary grass elevates its growing points, so frequent and close grazing may kill plants or thin the stand in the pasture.

Timothy

Timothy (*Phleum pratense* L.) is a somewhat shallow-rooted bunchgrass and is one of the most winter-hardy forage grasses. It does not do well in drought conditions. It is often grown with red clover or alfalfa for hay, and with careful management can also be used as part of a pasture mix. Several varieties of this grass are available; date to maturity differs by variety. It is worth researching which variety is best, depending on which legume types are grown with it.

Timothy has a bulb-like swollen base on the stem just above the soil level where some of the carbohydrates used for regrowth are stored. In addition, like brome and reed canary grass, it elongates its growing points during some times of the year, so it does best if grazed leaving a taller residual behind, and when given an extended regrowth period. For this reason, timothy is more frequently used as a hay species. However, with good grazing management where post-grazing residual height is not too short and regrowth periods are long, it can be an excellent part of the pasture mix. This species can be a good choice to include in a seeding mix on farms where the goal is to shift to a taller pregrazing height and taller post-grazing residual.

WARM-SEASON PERENNIAL GRASSES

As discussed in chapters 3 and 5, the warm-season grasses have their most productive season of growth during hot midsummer weather.

Bahia Grass

Bahia grass (*Paspalum notatum*) is mostly grown in warm, humid areas; it's one of the most common grasses used in Florida. It is a sod-forming grass and can outcompete other species in pasture once established. Leaves are low growing, so it is able to tolerate severe/short grazing. It is deep-rooted and drought-tolerant.

It can become invasive because it spreads both from seeds and rhizomes. However, it can be slow to establish because seeds are slow to germinate and seedlings don't quickly become vigorous. Thus, younger plants and seedlings will not compete well with weeds.

Bahia grass can grow on a wide range of soils including somewhat acidic and infertile soils. It does not do well on alkaline soils and won't tolerate shade. Yield is moderate on less fertile and marginal soils. The nutritive value of this grass declines as it matures, and it generally produces lower rates of gains for cattle when compared with other warm-season grasses.

Bermuda Grass

Bermuda grass (*Cynodon dactylon*) is a sod-forming grass that spreads by both stolons and rhizomes. This grass is mostly found in the warm, humid areas of the country. It is tolerant of short-term flooding but will not do well in soils that are poorly drained. It has good drought tolerance. Due to its ability to spread, it can fill in gaps in a pasture.

Much of the carbohydrate storage in Bermuda grass is found at or below soil level, so it can tolerate short grazing. It has only average nutritional value for livestock, but is still a choice for pastures due to its adaptability and persistence.

There are varieties that are grown from seed, and some hybrid cultivars that must be planted from vegetative material. The varieties that produce a lot of seed can become a weed species in some situations because they produce a significant amount of hard seeds, which will remain viable in the soil for many years.

Big Bluestem

Big bluestem (*Andropogon gerardi*) is a bunchgrass that grows 3 to 6 feet tall and is often red or purple when mature. This grass matures later in the growing season than switchgrass and maintains palatability better, even after seed heads form. It is adapted well to grazing because the growing points stay close to the ground until seed stems begin to form.

Big bluestem is more drought-tolerant than other warm-season grasses, so it can do well even in excessively drained soils. It is best adapted to deep fertile soils but can also grow in poorer soils. The seed is very light and it may be difficult to establish without good seed-to-soil contact at planting time.

Figure 6.5. Eastern gama grass being grown in a small demonstration planting shows its capacity to produce a significant amount of forage dry matter. Photo courtesy of Kevin Kaija, Vermont.

Eastern Gama Grass

Eastern gama grass (*Tripsacum dactyloides*) is a leafy bunchgrass which grows 6 to 12 feet tall and is related to corn. It can be very high yielding and provide higher quality forage than other warm-season grasses. This is one of the earliest warm-season grasses to begin growing in the spring.

It can be slow to establish and needs a midsummer rest period in order to maintain a healthy plant population in the pasture. It does best in deep soils that have good water-holding capacity. A challenge in planting this grass is that it has many dormant seeds. Dormant seeds will not immediately germinate, but the germination rate can be improved by stratifying the seeds (chilling the wet seeds) and then planting before the seeds dry out.

Indian Grass

Indian grass (*Sorghastrum nutans*) generally starts growing somewhat later than switchgrass and big bluestem. Indian grass spreads slowly from short rhizomes, and grows to be 3 to 6 feet tall. It provides good-quality forage during the summer and remains moderately palatable when seed heads appear.

The seed is light and can be difficult to get established unless good seed-to-soil contact is created when the seeds are planted. This grass can accumulate prussic acid early in the growing season, so it should not be grazed in the spring when it is less than 6 to 8 inches tall.

Johnsongrass

Most books include Johnsongrass (*Sorghum halepense*) in the section on weeds. Many areas prohibit the sale of Johnsongrass and list it as an invasive weed. However, it is already growing on many farms in the United States, and if managed well it can be a productive part of the pasture mix. If the goal is to eliminate Johnsongrass instead of managing it as part of the pasture mix, then grazing management can also be used to decrease the amount of this plant in the pasture.

Johnsongrass can be very competitive. It is fairly drought-tolerant and will often continue growing after other pasture grasses go dormant. During serious droughts, however, pastures that are primarily Johnsongrass may need to be avoided due to risk of toxicity from both prussic acid and nitrates. This is also true when grazing after an early frost. Because of these risks, many farms manage their pastures to eliminate this plant.

When managed carefully, Johnsongrass can be a relatively high-quality summer forage. Immature Johnsongrass is highly palatable, and if grazing livestock have access to it at this stage they can overgraze it and reduce or kill the plants in the pasture. The growing point of Johnsongrass is elevated 4 to 8 inches above the soil surface, so it is quite susceptible to severe and frequent grazing when it begins growing in the early summer or spring. This early-season severe grazing is one strategy to get rid of the plant in a pasture.

Johnsongrass can mature quite quickly, however, particularly when there is overgrazing or drought stress. Stems become thick and seed heads form quickly, and at this point the plant becomes significantly less palatable. If the goal is to maintain some Johnsongrass in the pasture and not kill it by overgrazing, livestock should be turned in when the grass is 12 to 18 inches tall. It should be grazed no shorter than 8 inches.

Figure 6.6. Johnsongrass is growing in this pasture along with tall fescue and several other species well adapted to hot, dry summers.

Little Bluestem

Little bluestem (*Schizachyrium scoparium*) is a smaller warm-season perennial bunchgrass with lower yields than some of the other species. It grows 2 to 4 feet tall and spreads by seeds, tillers, and short rhizomes.

It can grow on a wide range of soils, but does best on limestone-based soils. This grass is also less tolerant of wet soils and can be slightly less palatable than big bluestem, although it's also somewhat more tolerant of drought.

Switchgrass

Switchgrass (*Panicum virgatum*) is a tall rhizomatous perennial that will reach 3 to 5 feet tall. Although it may frequently grow like a bunch grass, the rhizomes form a sod, particularly under well-managed grazing. Basal

Figure 6.7. On this Pennsylvania farm, this switchgrass is thriving in a spot that receives longer rest periods between cutting and grazings than areas of the farm where cool-season grasses are grown.

tillers and shoots on the lower stems can produce late-season leafy growth. This warm-season grass includes varieties that are more cold-hardy than other warm-season grasses, so it can be grown in northern areas where the primary grazing species are usually cool-season perennials. It is also able to tolerate poorly drained soils better than other warm-season grasses.

It can be more easily established than big bluestem or Indian grass, although it's less palatable once seed heads appear. Switchgrass contains a chemical (diosgenin) that is toxic to nonruminants such as horses.

ANNUAL GRASSES

You may be surprised to learn that some grasses that reappear in pastures each year are annuals. These grasses reseed themselves each season. In addition to these "volunteer" annuals, there are several high-quality annual grasses that can be planted alone or as part of a mix for grazing.

Annual Bluegrass

Annual bluegrass (*Poa annua* L.) is generally considered a weed. However, this cool-season annual is commonly found in pastures and is sometimes mistaken for Kentucky bluegrass. It is often found in wet and compacted soils, such as high-traffic pasture areas or areas that have been overgrazed.

It grows in small tufts or clumps and usually has leaf blades that are shorter than those of Kentucky bluegrass. It also tends to grow earlier in the season than the other bluegrasses. Because it's fairly short, it is not high yielding.

Annual Ryegrass

Annual ryegrass (*Lolium multiflorum*) is a cool-season bunchgrass that can tolerate cool soil temperatures better than many other winter annuals. It is adapted to a wide range of soil and climate conditions but does best in cool, moist climates. It can tolerate wet soils and even short periods of flooding.

Seedling growth is very vigorous and competitive with the other pasture plants. Most of its growth is in late fall and early spring so it can be a useful species for extending the grazing season at both ends.

Barnyard Grass

Barnyard grass (*Echinochloa crus-galli*) is an annual warm-season grass that is commonly considered a weed. However, it can produce a moderate yield of pasture or hay when kept vegetative.

It produces large quantities of seeds, which provide excellent food for birds and wildlife but also make it difficult to eradicate the plant from pastures if it is allowed to go to seed. Managing barnyard grass by grazing it more frequently or grazing it shorter to keep it vegetative will make it more palatable and digestible, and prevent it from making seed.

Corn

Corn (*Zea mays*) is a high-yielding warm-season grass that requires a high level of soil fertility. It can be planted for harvest as a grain crop or silage, or it can be planted as a crop to be grazed. Corn varieties that have been selected for grazing are leafier, have smaller stalks, and are more digestible than standard silage or ear corn plants.

Corn does best in a crop rotation that maintains soil fertility and a weed management system based on prevention through rotation and cultivation. Grazing corn will be most successful if you use strip grazing to minimize trampling and waste. This will also limit intake so that other forages can be provided to allow a balanced diet to the grazing livestock.

Crabgrass

Crabgrass (*Digitaria* spp.) is a warm-season annual that is commonly considered a weed. However, it can provide palatable midsummer pasture if managed correctly. This grass can tolerate drought and heat well and is best adapted to well-drained soils.

Several improved varieties are available, but some farms find that good management of the common species already in the pasture results in palatable forage. Since it is an annual, crabgrass must come back from seed each year, so it must be rested in late summer to allow it to set seed. It can also be maintained in the pasture by being reseeded each year. It can be grazed as soon as 35 days after seeding, but may need as long as 60 days to establish.

The Foxtail Grasses

These warm-season annual bunchgrasses are found in pastures in many parts of the country. They are generally considered weeds, and are most common in moist soils. Giant foxtail (*Setaria faberi*) is the largest one, reaching 6 feet in some areas. Yellow foxtail (*S. pumil*) can be hard to differentiate from green foxtail (*S. viridis*). Look for long twisted hairs at the base of the leaf blade on yellow foxtail.

The foxtails produce a large amount of seed, which is somewhat millet-like, and can be an important food source for birds and wildlife. The prolific seed production also makes it difficult to eradicate foxtails from pastures. The leaves can be nutritive if they are in the vegetative stage, but in some situations can accumulate nitrates, which may create toxicity issues for livestock. They may also mature faster than other grass species in a pasture, becoming less digestible and reseeding themselves before other species have had time to regrow. The spiky seed heads can cause injuries to livestock including cuts, embedded seed heads, and mouth sores.

Millet

Millet (*Pennisetum americanum*) is a fast-growing, high-yielding warm-season crop. Pearl millet is frequently the preferred type for grazing, because it will regrow for multiple grazings during the summer. Millet provides a leafier alternative to grazing sorghum sudan grass. Another benefit is that millet does not carry the same risk of prussic acid poisoning as does sorghum sudan grass.

Pearl millet is better adapted to a range of soil conditions than sorghum sudan grass. However, it is more susceptible to cold stress, so it should be planted later once soils are warmer.

Wait to graze this annual until it is at least 18 to 24 inches tall, and leave a residual of 8 to 12 inches to allow for regrowth. This plant can regrow rapidly, so it can provide a significant amount of forage from multiple grazings during midsummer when perennial cool-season pastures are less productive and need a longer rest period. (See figures 5.15 and 5.16 for photos of millet.)

Small Grains

Small grains such as oats, barley, and triticale can be grazed when they are mostly vegetative. They can also be harvested as forage or allowed to mature and be harvested as a grain crop. They are generally a high-energy crop, either as a forage or grain. Their relatively short growing season makes it possible to mix them in with other crops in a rotation or plant them as a mix with a legume. They can be used for both midsummer and fall grazing.

Grains can be part of a mix of several species along with a legume when planted for grazing. Small grains include both winter grains and spring-planted grains. The winter grains are planted in the fall and overwintered for spring grazing. The spring grains are planted in the spring and grazed that same season.

Small grains may also be used as a nurse crop when planting a pasture mix. A nurse crop can help protect the new seedling plants as they get established and can also provide a higher forage yield during the first harvest.

Barley (*Hordeum vulgare*) has a bunch-type growth habit and fibrous roots. It can be grown as a grain crop, cover crop, or forage crop, or it can be grazed. It prefers well-drained soils and will not tolerate poorly drained soils. It is sensitive to low pH but can tolerate drier, less fertile conditions than wheat.

Oat (*Avena sativa*) is a palatable and productive bunch-type plant. Oats can provide either summer or fall forage for grazing and are frequently grown with field peas. Unlike cereal rye, triticale, or wheat, oat does not go dormant in the fall and will continue to grow even after light frosts.

Plants should be at least 8 to 10 inches tall before grazing begins. Digestibility and protein content will decline as the plants mature and begin to make seeds. Some varieties of forage oats are available, and they are leafier than standard grain varieties, with higher digestibility when grazed.

Rye (*Secale cereale*) is used for grain, grazing, or harvested forage. It is also used extensively as a winter cover crop. This cold-tolerant plant is well suited for fall planting to provide early-spring forage. Rye is more tolerant of a wide range of soil and environmental conditions than other small grains. Rye is not the same plant as annual ryegrass, which was discussed earlier in this chapter.

Wheat (*Triticum aestivum*) includes both spring wheat and winter wheat varieties. It is generally thought that rye or triticale are more productive for grazing. Wheat is also less tolerant of wet soils than rye or oats, and less tolerant of acidic soils than rye.

Sorghum Sudan Grass

Sorghum sudan grass (*Sorghum bicolor*) is a tall warm-season grass. It is a cross of two related grasses, sorghum and sudan grass. Sorghum is a tall coarse species grown for both grain and forage. Sudan grass has a finer stem than sorghum. Sorghum sudan grass has coarser stems than sudan grass, but also higher yields and a taller growth habit. This grass will regrow after grazing, and a number of commercial varieties are available. It should not be grazed shorter than 5 to 7 inches to allow for regrowth.

Figure 6.8. In this stand of sorghum sudan grass at the Horizon Organic dairy farm in Maryland, the contrast between the portion strip grazed by dairy cows and the intact plants demonstrates the large quantity of high-quality forage which can be supplied by this crop.

There is a risk of prussic acid poisoning if this grass is grazed when it is too young, frosted, or drought-stressed. To reduce this risk, it should not be grazed until it is at least 18 to 24 inches tall, and it should not be grazed after an early frost or during severe drought stress. It can also present a risk of nitrate poisoning in overfertilized stands during drought or stressful slow-growth periods.

Witchgrass

Witchgrass (*Panicum capillare*) is an annual warm-season grass found in many regions. It may appear as sprawling or tufted plants or grow upright to 1 to 2 feet tall. It is a common weed in poor-quality pastures and is of low nutritive value to livestock. Its seeds may serve as a food source for wild birds.

The Legumes

Legumes include both perennials and annuals, and most provide high-quality forage when managed correctly. As discussed in earlier chapters, legumes also fix nitrogen, which improves yields of other species in the pasture.

PERENNIAL LEGUMES

Perennial legumes include some such as white clover that will persist for many years in the pasture, and others that are shorter-lived. Including at least one perennial legume in any pasture mix is an important way to increase the nutritional content of the forage and provide nitrogen for the grasses and forbs in the pasture.

Alfalfa

Alfalfa (*Medicago sativa* L.) is a very popular deep-rooted legume. It is drought-tolerant due to those deep roots, but it may still go dormant during severe droughts. This is a high-yielding and highly nutritious forage plant. However, it is not as palatable as clovers. It may cause bloating in grazing livestock, particularly if the plants have been frosted. It is often grown with a grass such as timothy or orchard grass and harvested for hay or silage, but it may also be included as part of a pasture mix.

The length of time an alfalfa stand persists depends on climate, harvest management, soil fertility, and soil conditions. It tends not to persist very long in poorly drained acidic soils or under severe grazing. The crown of alfalfa plants is above the soil surface, so it can be damaged or removed by short grazing or by high-stock-density grazing with heavy trampling. It also does not tolerate frequent grazing. It does best when a taller post-grazing residual is left and regrowth periods are long.

It will do best in soils with a relatively neutral pH with good fertility. It is particularly sensitive to soil potassium, phosphorus, and boron deficiencies.

Bird's-Foot Trefoil

Bird's-foot trefoil (*Lotus corniculatus*) is a perennial legume well adapted to acidic and poorly drained soils. It has an intermediate growth habit, though there are some taller varieties that are better adapted for hay production. It persists longer than red clover in pastures and is well adapted to grazing. It needs a longer regrowth period than white clover so it may not persist under frequent grazings with shorter recovery periods.

This legume is best adapted to cool, humid regions, and can grow in a variety of soil conditions. It has limited tolerance to shading and drought, but will survive in wet

Figure 6.9. This field of alfalfa has low plant density, which is less than ideal for forage production or grazing, but lets us see the plants clearly. This low density occurred because the alfalfa was planted along with a grass species that was not cold-hardy enough to survive the winter.

Figure 6.10. Bird's-foot trefoil has an intermediate growth habit. It does not grow fully upright like red clover, but its growth is not as horizontal as white clover.

or waterlogged soils. It is most productive on fertile soils that are moderately to well drained. Bird's-foot trefoil, if allowed to go to seed, will reseed itself, although it's slow to establish in new pasture seedings. The plants generally live two to four years; then new plants will arise from either seeds or new shoots from existing plants.

Unlike the other legumes, bird's-foot trefoil does not cause bloat. This legume also contains condensed tannins, which can provide some help with the management of internal parasites.

In the fall this legume does not lose its leaves after a frost, making it well suited for fall stockpiling. It is similarly high in quality to alfalfa, but lower yielding.

Clovers

Alsike clover (*Trifolium hybridum*) is a short-lived perennial clover that will do well in poorly drained soils. It has a growth habit similar to that of red clover, with less rigid and upright stems. It can tolerate acidic, wet, and alkaline soils better than most clovers.

Unlike red and white clover, alsike clover doesn't tend to establish as well from frost seeding. It is suitable for grazing or haying, but because its stems are less rigid than alfalfa's it may tip over or lodge, making it difficult to harvest mechanically.

It produces high-quality forage, though it may cause photosensitization problems in some horses. Like many of the legumes, it can cause bloat.

Red clover (*Trifolium pratense*) is a short-lived clover, generally persisting for only a couple of years if it isn't frost seeded or allowed to go to seed. It is often grown in pastures and hay crops to improve forage quality and provide nitrogen, or as a green manure or cover crop. It is an upright growing plant with a taproot. It produces excellent-quality forage but, like many legumes, can cause bloat.

Figure 6.11. Red clover is the primary legume in this recently planted high-quality pasture in northern Vermont. Red clover has an upright growth habit, so it's well suited for the diverse mix of tall, leafy, cool-season grasses in this pasture.

Figure 6.12. White clover can mix well with cool-season grass species and create a dense sod.

Although it has deeper roots than white clover, it is not as drought-tolerant as alfalfa. It is only moderately tolerant of wet soils. It is more tolerant of acidic soils, less fertile soils, and shade than alfalfa.

It tends not to persist as well as white clover does under short or frequent grazing, but with occasional frost seeding into a grass mix it can produce high yields. It will also reseed itself if allowed to go to seed in the pasture. Red clover seed tends to be relatively inexpensive, so regularly adding it to pastures in frost seedings can be a low-cost investment that pays off well in improved pasture quality.

Subterranean clover (*Trifolium subterraneum* L.) is a reseeding legume that is well adapted to grazing. A low-growing plant that forms a dense sod, it's a palatable pasture species and will grow in non-irrigated pastures under mild drought conditions.

It does not tolerate shade well but can grow in somewhat acidic and poorly drained soils. There are many varieties; most do well in regions with winters that are moist and mild and drier summers.

White clover (*Trifolium repens* L.) is a shallow-rooted legume well adapted to grazing. It is widely adapted and found throughout the United States and Canada. It is winter-hardy, and grows best under moist, cool conditions.

White clover includes low-growing "wild" varieties found in perennial pastures, and taller "ladino"-type clovers found in intensively managed pastures and some hay fields. It is well adapted to grazing, and spreads naturally under good grazing management. It mixes well with cool-season perennial grasses in a pasture mix.

White clover stolons grow horizontally over the surface of the soil, so when grazed, only leaves, petioles, and flowers are removed. This allows the plant to be grazed without having any growing points removed, which is one of the growth habits that makes it so well adapted to grazing. The feed quality is also very high, as the less digestible stems are not consumed.

Because of its horizontal growth habit, a single white clover plant may cover a large area in the pasture. Thus, different parts of the same plant may experience different growing conditions and respond with different growth habits. For example, some parts of a plant may be exposed to sun while other parts are in the shady pasture canopy. Since roots, as well as new leaves, form at nodes in the stolons, this single large plant can become two new plants if separated from the original "parent" plant.

White clover can be frost seeded into a pasture to improve forage quality. White clover is a hard-coated seed that frost seeds well, and under certain conditions seed will sit dormant for many years before germinating.

Like many legumes, white clover can cause bloating in cattle and sheep under some conditions.

Sericea Lespedeza

Sericea lespedeza (*Lespedeza cuneata*) is a perennial legume that grows in warmer regions; it is not able to tolerate cold northern climates. There are some annual varieties of lespedeza. Perennial types are deep-rooted, long-lived, and drought-tolerant. This legume grows upright from 2 to 4 feet tall. It will grow on infertile and acidic soils but is much more productive on fertile soils. It can grow in soils that many other legumes will not survive in, and in some areas it is considered an invasive weed instead of a beneficial pasture legume.

It produces medium-quality forage, but can become less digestible as it matures and produces more stem. Quality will be higher if it's grazed when vegetative. This is one of the non-bloat-causing legumes. It also contains condensed tannins; like bird's-foot trefoil, it may help with the management of gastrointestinal parasites.

Vetch

Hairy vetch (*Vicia villosa*) is a short-lived perennial that can grow in a range of soil conditions. It has a short vining growth habit and can be either grazed or harvested as part of a hay crop mix. It may also be mixed with annuals such as oats, peas, and vetch for forage or pasture. If you're grazing this plant, it does best if it is not grazed short. This is the most cold-hardy vetch, and is more tolerant of poorly drained soils than common vetch.

Common vetch (*Vicia sativa*) is less winter-hardy than hairy vetch, and commonly used as a cover crop. It can also be used for grazing or hay production.

ANNUAL LEGUMES

Cowpea (*Vigna unguiculata*) is an annual legume that can be included in a pasture mixture with other annuals such as millet, small grains, or sorghum sudan grass. Botanically speaking, this legume is actually a bean, even though it's called a pea. It will do best in well-drained soils but can tolerate a variety of soil fertility and somewhat acidic soils.

Field pea (*Pisum sativum*) is a highly productive annual legume that does well in cool, humid conditions. This forage crop does well in a variety of soil conditions but won't tolerate drought well. This legume is high in protein and highly digestible. For grazing, field peas are usually mixed with cereal grains such as oats or triticale to provide a broader range of forage nutrition, and to help hold up their vining stems and leaves.

Figure 6.13. Peas, grown here in a mix with small grains and brassicas, serve both as a cover crop for a vegetable crop rotation and as late-summer pasture for the herd on the Flack Family Farm in Vermont.

Lablab (*Lablab purpureus*) — usually grown as an annual legume — is similar to a large cowpea. It is high in protein and highly digestible. This legume is fairly tolerant of heat and drought but will not tolerate poorly drained soils. It grows well in a mix with tall annuals such as millet or sorghum sudan grass.

Nonlegume Forbs

Pastures may contain other potentially beneficial plants that are neither a legume nor a grass. Some of these forbs may be planted, while others will appear naturally under good pasture management.

BRASSICAS

Brassicas include forage varieties of kale and turnips. These high-yielding, highly digestible crops can be used for summer or fall grazing.

These plants are high in protein and energy, so they can be a great addition to the overall ration. However, they are very low in fiber so they must be fed as just part of the pasture mix or feed ration. Brassicas can cause health problems if not grazed correctly and carefully. Problems from feeding too much brassica can include bloat, nitrate poisoning, atypical pneumonia, hypothyroidism, and hemolytic anemia.

To prevent these problems, introduce grazing animals to brassicas slowly, generally over four to five days. Make sure the herd or flock is not hungry when turned into the brassica field. Strip grazing is a common way to carefully graze brassicas and limit daily intake by livestock. Brassicas should comprise only a portion of their feed.

Brassicas can be used for fall grazing to extend the grazing season, but some varieties will need 80 to 150 days of growth after seeding. Brassicas are not drought-tolerant and do best in soils with moderate to high fertility. They may be grown in a mix with other annuals or drilled into an existing grass or legume crop.

Kale

Kale (*Brassica oleracea*) is one of the more cold- and frost-tolerant brassicas. Livestock graze the leaves and stems. Most kale varieties require 150 or more days from seeding to grazing in most areas. Quite a range of kale varieties exist, including stemless varieties, taller or shorter varieties, and a few varieties that are able to grow rapidly to allow grazing sooner. Kale can generally only be grazed once, although some stemless varieties can be grazed twice.

Rape

Rape (*Brassica napus*) is a multistemmed type of brassica, and livestock will eat both stems and leaves. This plant may regrow after grazing, and requires less time to grow after seeding than kale to be ready for grazing. Some varieties mature sooner than others, and some are less suited for grazing and fall stockpiling than others. Varieties also range in height, so choose one that will work well with the type of pasture mix you are growing. Some types of rape are grown as an oilseed crop (canola).

In order to get two grazings from this crop, leave a stubble of 6 to 10 inches after the first grazing. Resume grazing in about four weeks; on the final grazing, let the animals graze the plants short.

Turnip

Turnip (*Brassica rapa*), like rape, can be a good fall grazing crop. Livestock graze the leaves, stems, and aboveground root bulbs. Turnips can be grazed twice if the first grazing removes only the leaves, which allows the stand to regrow from the tops of the roots. A second grazing can be done after about four weeks of regrowth in good fall growing conditions.

DANDELION

Dandelions (*Taraxacum officinale*) are not usually intentionally planted in a pasture, but they will likely appear anyway. This taprooted plant is palatable and provides a high forage quality when vegetative. It's widely adapted and is found throughout the United States. Including some plants like dandelion that have a significant taproot can help improve soil due to improved diversity of rooting systems and more roots reaching more areas in the soil. In pastures that have suffered overgrazing damage, however, dandelion can become overabundant and be considered a weed species.

Forage Chicory

Forage chicory (*Cichorium intybus*) is a somewhat short-lived perennial with a deep taproot. Plants can produce highly digestible forage with a high mineral content. The taproot gives it some drought tolerance. The leaves emerge as a rosette, and the plants look somewhat like a dandelion, though larger and taller.

It produces lush leaf growth in the spring and throughout the first year it is planted. Beginning in the second year, a flower stem will emerge in late spring. This can reach 5 to 6 feet tall with a thick stem, which has lower palatability and digestibility. The flowers are blue and quite similar to the common roadside chicory, which is the same species. To prevent bolting and maintain more vegetative digestible forage, graze or clip the stems when they are 6 to 10 inches tall.

Good grazing management is required to maintain this plant for five to seven years in the pasture. The crown and taproot of chicory can be damaged by over-grazing, severe short grazing, or excessive trampling. It is particularly sensitive to trampling when newly seeded.

Chicory does best in fertile, well-drained soils, but it can tolerate low pH. It can be planted as part of a pasture mix, or it can be drilled or broadcast into an existing pasture. If broadcast seeding is done, spreading the seed in the spring will create better seed-to-soil contact.

Chicory needs 80 to 100 days to establish before it is ready to be grazed. Once established, it's a fast-growing plant, so if it is strip grazed without a back fence, livestock may be able to graze regrowth and do overgrazing damage.

In addition to being highly digestible and high in minerals, chicory contains condensed tannins and sesquiterpene lactones, which may help with the management of gastrointestinal parasites.

Plantain

Plantain (*Plantago lanceolata*) is often thought to be just a pasture weed, but this perennial plant is actually a highly palatable and nutritious addition to pasture diversity. There are two types of plantain that may appear in pastures: the broad-leafed variety (*P. major*) and the narrow-leafed (*P. lanceolata*) type. The latter is the one that is now commercially available to include in pasture seed mixes.

Figure 6.14. Spring Creek Farms in Pennsylvania grows forage chicory in a mix with several grass species, alfalfa, and red clover.

Figure 6.15. Narrow-leafed plantain is highly palatable and adds diversity to the pasture.

Plantain leaves are similar to chicory in nutritional content, but older leaves and seed stems become unpalatable, with lower digestibility. Plantain establishes rapidly in pastures and will persist in a wide range of soil conditions.

OTHER FORBS AND WEEDS

Other forbs/weeds in pastures include a large number of broad-leafed plants that may be woody (browse) or vegetative. Some (such as thistles) may not be very palatable; a few may even be toxic. However, even those considered weeds may not be a problem when found growing only in small amounts, and can contribute to the overall nutrition of the forage in the pasture.

When a forb or weed begins to spread and increase in numbers in a pasture, then some attention should be paid to the possible cause, and whether the plant poses any risk of becoming an invasive species. If a species appears to be spreading and causing problems in the pasture, then it is likely that the cause is poor grazing management practices. Refer to chapters 3 and 5 for information on how to improve grazing practices to prevent weeds from becoming a problem.

It is also important with some forb species to consider whether there may be problems with toxicity or whether the presence of the plants may lower livestock pasture dry matter intake. See chapter 12 for more information on potentially toxic plants.

The Art of Good Grazing
Raising Beef, Improving Ecosystem Health

Tussock Sedge Farm in eastern Pennsylvania is owned and run by Henry and Charlotte Rosenberger with help from their farm manager, Jarett E. Brown, and Megan Nesrud, who handles marketing. The farm is just under 600 acres and consists of multiple tracts of land surrounding the local town. More than half of the farmland is now preserved permanently as open agricultural land.

In addition to farmland preservation, Henry and Charlotte have also restored almost 30 acres of wetland on the farm. The farm has a high diversity of wetland plants, shrubs, and grasses; they have planted over 1,000 native trees. The restored wetland includes ponds, streams, dams, and other natural structures to slow and capture water as it flows through the farm. During some high-rainfall events, these areas have played an important role in preventing flood damage on the farm and in downstream communities. There is an abundant wildlife population, including many bird species, thanks to improvements they have made to habitat. In addition, nearly 60 birdhouses have been placed on the farm, built mostly by local Girl Scouts. Local birders who enjoy

bird-watching on the farm maintain these. A recent bird survey on the farm found that over 150 different species were present! Figure 4.4 shows one of the birdhouses on this farm; it is mounted on a fence post along the edge of a protected riparian area.

The beef herd, which consists of Red Angus and Devons, is divided into smaller herds. This allows the

Figure 6.16. At Tussock Sedge Farm, livestock have access to a mineral feeder, which is moved from paddock to paddock with the herd.

Figure 6.17. Polywire is used to subdivide the pastures into smaller paddocks. The polywire reel is hung on the high-tensile fence, which functions as an end post to keep the polywire fence tight and also provides power to electrify the polywire.

Figure 6.18. The water system on the farm is 180 psi 1-inch pipe, which is mostly run along the lane fences or perimeter fences.

nutritional needs of each group to be managed separately. It also allows the farmers and staff to manage these smaller cattle groups on separate tracts of land to reduce the need to truck cattle from one area to another.

Henry and Charlotte have worked hard to improve the cattle genetics on the farm so the animals are able to grow and finish well on 100 percent forages. Cattle are fed only pasture during the grazing season; in the winter they are fed hay and balage and kept off pastures to avoid damage to plants and soils.

Demand for their beef has grown, and Tussock Sedge now direct markets a significant amount of meat from an on-farm store. There are other local markets as well, so only a small amount of meat is sold through wholesale channels. As interest in the environmental and human health benefits of grassfed meats has grown, Henry and Charlotte have found there is now market demand for the organ meats, bones, and other "parts" that were previously hard to sell. Henry is exploring ways to grow the herd so that the farm can keep up with the rapidly increasing market demand for its beef. Based on his experience with cattle in the past, though, Henry says he is only willing to graze cattle that have the right genetics to produce high-quality beef from 100 percent forages.

The four different grazing groups include an older finishing group, a younger growing group, a spring calving group of cows, and a fall calving group. Each individual property of the farm has its own grazing rotation with one group of cattle. Fencing is high tensile on the perimeters and lanes, with a single strand of polywire used to subdivide each area into smaller paddocks. Lanes make it possible to move groups of animals around each property without having them walk through regrowing pastures and hay fields. Much of the fencing on the farm was built with assistance from NRCS (Natural Resources Conservation Service) programs.

Cattle are moved to fresh pasture every one to three days. Water is piped to each area for drinking so the cattle don't have to walk back through recovering pasture to drink. Portable water tubs with float valves are moved with the herd and connected to the water line to keep them full.

The farmers have paid a lot of attention to soil health and test soils in each field every three years. With an emphasis on good grazing practices and the use of composted winter manure as a soil amendment, they have built soil organic matter content from little or none to 4 to 7 percent in most fields. As the soil organic matter content and soil fertility have improved, they have found that plants and soils are less impacted by drought and other climate-related challenges. This is allowing the farmers to transition the pasture species mixture to more palatable and nutritious legumes and grasses. They do some frost seeding of both red and white clover most years.

The specific length of time cattle graze an area depends on the time of year and pasture growth. After each grazing, the fields are left to regrow for as little as three weeks (when plant growth is rapid in the spring) to as much as six weeks (later in the year when pasture growth slows). By varying the stocking density at different times of the year, the farmers achieve a different trampling effect. Areas that were more heavily trampled, with a higher stock density, don't need clipping as often. Paddocks are clipped after the second or third rotation if needed to remove tall fescue seed heads or weeds. Whether paddocks are trampled or clipped, they regrow an equally high-quality, highly digestible vegetative crop of forage.

Figure 6.19. The left side of this pasture looks like it was mechanically clipped after the previous grazing. However, that area was actually just grazed with such a high stock density that the grass seed heads were trampled.

PART THREE

Grazing from the Animal's Perspective

Ruminants (cows, sheep, and goats) have digestive systems that, with assistance from an amazing diversity of microorganisms, have the ability to transform high-fiber forages (which humans can't digest) into meat, milk, and fiber.

Much of the land in North America and other parts of the world is not suited for annual arable crop production for human consumption. But when farmers work with ruminant livestock and thoughtfully planned grazing systems, those areas can often be sustainably managed in perennial forages, on which ruminants can thrive. Thus, through good grazing management, these non-arable lands can become sources of food, clothing, income, and improved soil and ecosystem health.

However, just as pasture plants have needs that must be met in order to thrive and produce high-quality feed, grazing livestock must also have their basic needs met. A good understanding of ruminant anatomy, function, and needs is helpful in creating sustainable ecologically sound, grass-based livestock farms. By learning more about ruminants' grazing behavior — along with their nutritional, environmental, and social needs — we will better understand why livestock choose to eat certain plants and trample others. This knowledge will help assure we provide the correct amounts and types of supplements when needed, and that we are able to monitor the health and well-being of the herd or flock. The more knowledge we gain of both plant and animal needs, the easier it becomes to create grazing systems that humanely care for the livestock while also employing them as a natural force that can continue to improve pasture quality and soil health.

CHAPTER 7

The Unique Digestive Ability of Ruminants

Ruminants include sheep, goats, and cows. They are hoofed mammals whose unique digestive system makes it possible for them to digest the fibers in plants that humans and other animals can't. Unlike monogastric (one-stomach) animals such as poultry and pigs, ruminants are able to digest and utilize fibrous forages, including pasture.

But ruminants don't actually do this digestive work all by themselves. Cows, goats, and sheep don't have the enzymes necessary to digest plant fibers directly. Instead, they and other ruminants rely on a symbiotic relationship with microorganisms that live in their digestive system and are able to break down the fibers in the plants and make the nutrients available. With the assistance of these microorganisms, ruminants can utilize forage by fermenting it in some of the multiple compartments of their digestive system.

This relationship of livestock to microorganisms is parallel to the dependence of plants on soil microorganisms to make nutrients available. Thus, while grass farmers may be growing and selling meat or dairy products, and calling themselves grass farmers, they could also accurately call themselves microbe farmers. Soil microbes and gut microbes make possible much of the conversion of sunlight-fueled forages into livestock products.

Ruminant Anatomy

Ruminant digestive systems are multi-compartmented, with four "stomachs." This is different from other animals (including humans) that have simple monogastric (single-stomach) digestive systems. These four compartments each have a specialized function in the process of digestion. The compartments include the rumen (from which ruminants get their name) along with the the reticulum, omasum, and abomasum.

This four-part "stomach" takes up about three-quarters of the space inside the abdomen of the animal, as shown in figure 7.1. The rumen and reticulum make up the largest part of the total stomach volume, and in a mature cow, these may have a capacity of over 40 gallons. The rumen and reticulum are not actually separate compartments, as they are separated from one another by only a fold of tissue. These two compartments are often called the rumino-reticulum. More of the rumen volume is located on the left side of the animal, which makes it possible to observe how full the rumen is by observing the animal's left flank, as shown in figure 7.2. This is a very helpful way to determine if an animal is getting enough forage intake from pasture.

The digestive system of a ruminant starts at the mouth, where saliva is added to the forage to help with digestion and swallowing. The saliva contains enzymes that break down some of the fats and starches. But saliva also helps to buffer the pH of the digestive system. A mature cow may produce as much as 50 quarts of saliva per day.[1] This large quantity is necessary for digestion, and is one reason it's important that livestock always have access to good-quality drinking water.

From the mouth, the bolus of forage, which is now mixed with saliva, moves through the esophagus to the rumino-reticulum. Muscle contractions in the esophagus will later make it possible for a bolus, or cud, to move back upward to the mouth for further chewing if needed. This allows the animal to quickly fill its rumen

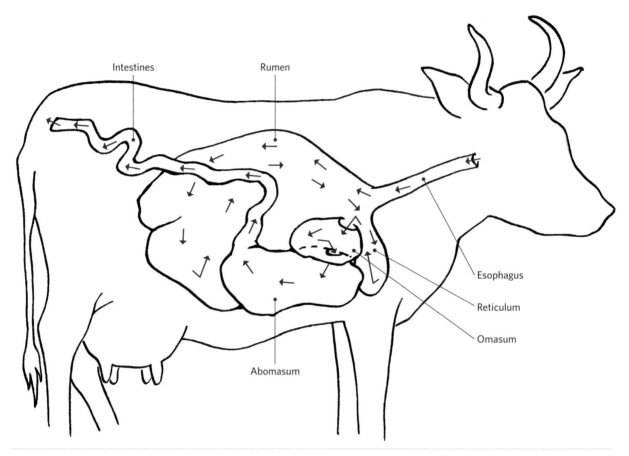

Figure 7.1. The four parts that make up the center of a ruminant's digestive system are the rumen, reticulum, omasum, and abomasum. Illustration by Anna Powell.

with pasture, and then chew more later, once it is in a comfortable safe place.

Only a small fold of tissue separates the first two stomachs, the rumen and reticulum, so partly digested material (digesta) and microbes can flow freely between them. A lot of mixing and churning happens during this part of digestion! Together the rumen and reticulum work as a large fermentation vat that contains a huge number of microorganisms.

The rumen microbes provide the needed digestive enzymes for the breakdown of plant fibers. In turn, the microbes get what nutrients they need for growth and reproduction. These organisms ferment and break down the fibrous cell walls into carbohydrates; they also produce fatty acids, which are an essential energy source for the animals. The organisms also synthesize proteins and vitamins,

which provide nutrition to the animals as well. Some microbes will pass from the rumen and reticulum farther along the digestive tract, where they are digested, providing still more nutrients. The relationship between the livestock and these microbes is a mutually beneficial one.

Note that what the animals are fed has a large effect on the type and health of organisms in the rumen. Sudden shifts in diet or overfeeding of some types of feed (such as grain) can result in weakened microbial populations and sick cows.

In addition to microbial fermentation at this stage of digestion, fiber digestion is also improved by cud chewing. Cud chewing grinds up cell walls to release the contents of the cells and break up more of the plant fibers. This allows the microorganisms to break down the fiber and utilize the cell contents more easily.

Figure 7.2. The filled-out triangle of soft tissue just under the short ribs of this cow indicates how full her rumen is.

From the rumino-reticulum, the smaller particles and liquids move into the omasum. The omasum has many folds of tissue. This is where much of the water is removed from the digesta. From the omasum, the digesta passes to the abomasum. The abomasum functions more like our own stomachs, using hydrochloric acid and enzymes to continue the digestive process.

Some types of ruminants, such as goats, are better browsers, while others (including cows) are better grazers. The grazers tend to have a larger rumen and reticulum, and the opening to the omasum is smaller. The browsers have a larger opening to the omasum, which allows a faster rate of passage. This may be an evolutionary adaptation that allows cows to leave the more digestible pasture forage in the rumen longer so that it has more time for more complete microbial fermentation. The browsers instead can more quickly pass the nondigestible fibers (lignin) common in browse plants out through the digestive system. This allows browsers like goats to move digesta through faster so they can have a higher intake. Being able to eat more, and pass it through their digestive system faster, helps make up for the lower digestibility of some of the plants they select and eat.[2]

Whatever the type of ruminant, from the abomasum, the digesta moves into the small intestine, where many of the nutrients are absorbed. In a mature cow, the small intestine has a capacity of 20 gallons, and may be 150 feet long. At the point where the small intestine meets the large intestine is the caecum. This pouch-like organ is where some secondary fermentation can occur before the digesta moves into the large intestine. More water is removed in the large intestine before excretion.

Why did this group of animals evolve with such a complicated and interesting digestive system? Bill Murphy, in *Greener Pastures on Your Side of the Fence*, wrote that "the grazing animal's lifestyle may have had a lot to do with how things turned out. The defenseless grazers probably would leave the forest cover only to fill up quickly on plants growing in clearings near the forest edge. Once full, they would return to hiding places to digest the food."[3]

Eat Now, Chew Later

Ruminants start the digestion process by harvesting mouthfuls of pasture and swallowing. This grazing or harvesting of the pasture is not done by "biting" off the pasture plants the way horses do. Horses have lower and upper teeth in the front of their jaws, so they are actually able to bite the plants off between their teeth. Ruminants have a lower row of incisor teeth, but instead of upper incisors, they have a muscular pad. They take forage into their mouths, then break off the mouthful by holding it between their teeth and the upper dental pad and tearing it off.

Cows, sheep, and goats each ingest mouthfuls in slightly different ways. Cows are better adapted to grazing leafy pasture plants; goats are better at browsing leaves of woody plants and brush. Sheep are mixed feeders, good at either grazing or browsing. This is due at least in part to some anatomical and physiological differences among these animals.

Cows have wider muzzles, teeth that point forward a bit like a spatula or scoop, and lips that are stiffer than a goat's. This allows cows to use their long dexterous tongues to reach out, pull the pasture plants into their mouths, and efficiently take big mouthfuls to fill their rumens as fast as possible. Due to the size and design of the cow's mouth, in a well-managed system she won't graze the pasture much shorter than about 2 inches above the soil. When grazing in an area of mostly browse species (bushes and small trees) rather than a more uniform pasture, cows will have a more difficult time than a goat selecting the most nutritious individual leaves.[4]

Browsing animals, such as goats, have more flexible lips, narrower muzzles, and more upright teeth in the front of their mouths. Goats have incredibly mobile

Figure 7.3. At Does' Leap goat dairy in northern Vermont, this doe is using her impressive selective grazing skills to eat leaves from the most palatable shrubs in a diverse pasture.

upper lips, which allow them to be picky or selective grazers. They can eat individual leaves off a shrub, even avoiding thorns as they consume the flowers and leaves of a plant such as a multiflora rose or hawthorn tree.

Goats are much more efficient browsers than cows or sheep and can defoliate young trees and brush rapidly, particularly palatable species. Goats can be even more selective than sheep in their grazing, particularly at low stocking densities. If pastures are managed with larger paddock sizes, and the goat flock is left in each paddock for a week or more, the result can be significant overgrazing damage of their favorite plants.

Sheep are also browsers, but are somewhere in between goats and cows in their grazing abilities and preferences. Like goats, they have the ability to use their lips and teeth to selectively graze pasture or browse brushy plants. In a pasture setting, their mouth design allows them to be more efficient browsers than cows, but it also allows them to graze pasture plants shorter than cows normally do. So when poorly managed with periods of occupation that are too long, sheep can damage

Figure 7.4. This ewe is pushing aside the nonpalatable buttercup to selectively graze the shorter cool-season grasses and clovers in the pasture.

pasture plants, particularly those with crowns and growing points above the soil surface.

Watching a herd or flock when the animals are first turned out into new pasture is a great way to observe their grazing behavior. As the animals walk forward, they will pull a certain amount of pasture growth into their mouth, then break it off and swallow it. If the pasture is fairly vegetative, they won't spend much time chewing at this point; they will concentrate on harvesting to fill their rumens with forage.

If the pasture is uniform and high-quality, there will be a steady rhythm to the grazing. The cow steps forward, grabs some forage with her tongue, tears it off, swallows, grabs some more forage, tears it off, swallows, and then steps forward to start the process over. Sheep and goats move similarly, but use their lips and teeth more than their tongues.

The sound of livestock grabbing mouthfuls of pasture and breathing adds to the meditative enjoyment of watching this process. But the observation also provides important information that can help you learn how to make it easier for livestock to efficiently harvest as much pasture as possible.

The grazing or forage harvesting process has been studied extensively. So we know how many bites various types of livestock can take per minute and per day, what the bite size is, and how much of the day they are able to spend grazing.

Cattle take an average of 25,000 to over 30,000 bites per day when grazing. Sheep may take over 50,000 bites per day.[5] But ruminants can't spend all of their waking moments wandering around in search of pasture. They also need time to rest, chew their cuds, drink some water, and engage in social activities with the rest of the herd or flock. The more we understand how much time and effort they put into grazing, the better we can design grazing systems that allow them to maximize dry matter intake from pasture as easily as possible.

The amount of intake from pasture can be determined by this formula:

$$\text{Grazing time per day} \times \text{biting rate} \times \text{amount of intake per bite}$$

What this means in its simplest interpretation is that the larger the bite size, the fewer bites animals have to take to meet their dry matter intake needs. So the more forage they can harvest per bite, the easier we make it for them. In addition, the higher the quality of the dry matter they get in each bite, the easier it is for them to do the pasture harvesting work for us.

Conversely, moving livestock into a pasture where the plants are very short limits their dry matter intake and productivity potential. So grass farmers want to make sure that the management choices they make ensure that each bite the animal takes includes as much high-quality dry matter as possible.

Farmers can help the livestock fill their rumens most efficiently on pasture by providing them with pastures that are managed to have a high density of plants that are palatable and digestible, and are the correct height at the time the animals are introduced there to graze. Chapter 9 covers this subject of managing to maximize pasture dry matter intake in detail.

The Art of Good Grazing
Custom-Grazing Dairy Heifers

Troy Bishopp is a fifth-generation farmer on the Bishopp Family Farm, located in central New York State. He also works as a grazing specialist for the Madison County Soil and Water Conservation District.

Instead of owning his own herd of cattle, Troy custom-grazes dairy heifers and beef cattle from other farms. At the end of the grazing season, he returns the livestock to their owners and prepares to start grazing a new group during the next season. The owners of the heifers are happy because the animals are healthy and grow well on the high-quality pasture. Troy is happy because he earns income (paid per head per day) without as much market risk, and he is able to steadily improve his land through good grazing management of the heifers and the pastures.

Troy's style of grazing management is carefully planned and has the objectives of making money and improving the soils, plants, water quality, and overall ecosystem health, while also providing high-quality feed for livestock. This good management has resulted in pastures that are full of a diverse mixture of deep-rooted perennial plants. The soils have improved, which is evident in the health of the plants and the high level of biological activity in the soils. The length of the grazing season (260 days) has also increased, which allows him to have an income from custom-grazing for a longer season.

Troy uses a grazing chart to both plan out each grazing season, and keep records of where and for how long the herd grazes. This planning process allows him to assure that each pasture is given a long regrowth period, which keeps the plants (and soil) as healthy as possible while also meeting his nutrient management goals. He also uses the chart to plan where to graze each group of heifers during the spring, summer, fall, and even in the winter months. Through his records, Troy has found that the right time to stop grazing on pastures where he wants growth to accumulate for stockpile grazing is at least 60 days before the first killing frost.

The Bishopps don't make hay on the farm, so they have to use livestock to harvest the pastures. Another benefit of their grazing planning is that it allows them to anticipate the excess forage, so that it can be more easily managed through grazing, with some post-grazing clipping when needed. This is done by moving the herd rapidly during times of year when the cool-season

Figure 7.5. Pastures on the farm include a diverse mix of high-quality cool-season grasses and legumes. The water tank located on this fence line is part of the system that provides gravity-powered drinking water to most of the pastures.

Figure 7.6. Strip grazing prevents these heifers from trampling snow-covered stockpiled pasture. Each forward move of the front polywire fence provides the herd with a long, narrow strip of standing forage. Photo courtesy of Troy Bishopp, Bishopp Family Farm.

perennial plants are growing rapidly and by using portable fencing to adjust paddock sizes and stock density. As the heifers move through those taller spring and summer pastures, they eat the most palatable and digestible plants. They leave behind plenty of leaves, trampled plants, and manure, which all help to improve both soil and plant health.

At any given day during the grazing season, there will be pastures at all stages of growth on the farm. Standing at the top of the hill provides a view that includes some pastures just beginning to regrow after being grazed, some pastures in intermediate stages of regrowth, and some that are ready for grazing.

Troy has planted hedgerows and restored a wetland on the farm. The trees in the hedgerows are protected from livestock grazing by fences, and have now grown enough that they are starting to provide shade and wind protection for the animals. In addition, they have created wildlife corridors through the farm for birds and other animals. There are also birdhouses located on fence posts throughout the farm to encourage nesting fly-catching species. These help keep down the fly population on the heifers.

Permanent electric fence is installed around the perimeter of the farm and between each main field. During some times of the year Troy uses a high stock density to create a more thorough post-grazing trampling effect in the pastures. But at other times it makes more sense to graze with a lower stock density. The use of portable posts and polywire to subdivide the pastures

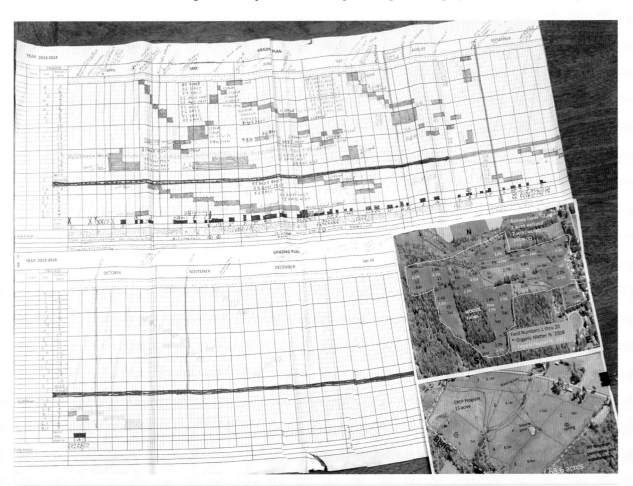

Figure 7.7. Troy keeps his grazing chart and the maps of the farm handy so he can refer to them regularly. See figure 10.2 for a closer look at the detail in one of these charts. Photo courtesy of Troy Bishopp, Bishopp Family Farm.

Figure 7.8. The hedgerows are protected by fence so the heifers don't eat the young trees.

allows the paddock sizes to be changed frequently. This makes it easy to quickly adjust the stocking density in each paddock.

Tubs with float valves are located throughout the farm to provide water for the cattle. Troy constructed a water catchment in the stream that runs through the property. Water is piped from the stream to the tubs. This avoids the need for the cattle to drink directly from the stream, which fits in with the farm's goal of improving riparian areas and protecting water quality.

CHAPTER 8

Ruminant Nutrition from Pasture

The basic nutritional requirements of all ruminants are similar: protein, energy, fiber, minerals, vitamins, and water. To get enough, they require a certain amount of daily dry matter intake (DMI) that has in it the right balance of these nutrients. What constitutes the "right" balance and amounts of nutrients varies, however, depending on the type of animal and other factors including body weight, genetics, environment, stage of life, and level of production.

Pasture forages contain different amounts of the required nutrition and also vary widely in how digestible they are. So while some forages are highly digestible, making it easy for livestock to meet their nutritional needs, others may be low quality, resulting in undernourished livestock. In addition to understanding the quantity of feed and the balance of nutrients the livestock need, we also need to be able to know how to measure quality using forage tests.

Once we know the animals' specific nutritional requirements, and what quality of forage is available, we can create a grazing system that maximizes dry matter intake from pasture, provides the right supplements, and meets the nutritional needs of the livestock. This is the ultimate goal, because it benefits the livestock and the bottom line: The more pasture they eat, the lower the feed costs, and the better the animal performance will be.

Basic Ruminant Nutrition

Ruminants need the right amount of protein, energy, fiber, minerals, vitamins, and water. Most of these essential nutrients can be provided from pasture plants and stored forages such as hay or silage, with some additional mineral supplementation and access to clean drinking water. But some farms may also feed grain to livestock. Let's look at each of these nutrients and what role they play in keeping the animals healthy and productive.

FIBER

Fiber is essential in the diet to keep the rumen functioning correctly. When fiber is in short supply, the rumen and the essential rumen microorganisms will stop working correctly, which can cause serious health problems. Insufficient fiber results in less saliva to provide a pH buffer in the rumen so the rumen will become more acidic. Lack of healthy fiber digestion will also decrease butterfat production.

On a grass farm with enough high-quality pasture, the herd or flock should be able to ingest sufficient fiber from the pasture. If the pastures are very lush at some times of the year and forage fiber levels are low, it may be beneficial to supplement with additional fiber from dry hay, beet pulp, balage, or another source.

The fiber in pasture plants is found in the plant cell walls. The cell walls are made up of cellulose, hemicellulose, and lignin. Cellulose is easily digested in the rumen by microbes. Hemicellulose is a more complex molecule, so it is less easily digested. Lignin is largely indigestible even by ruminants and their microorganism helpers. The more digestible forms of fiber help provide energy to the livestock, but even the nondigestible fiber plays an important role in digestion.

Energy is essential for growth and development, and animals require a lot more energy than protein each day. Energy in pasture forages comes from the soluble energy in the cell contents, as well as from the digestion of the fiber in the cell walls. Plants that are less mature contain less lignified fiber and more soluble and digestible fiber than fully mature pasture plants. The higher the amount of digestible fiber, such as cellulose, in the plants, the more energy the livestock will be able to glean from those plants.

It is important to note that the levels of energy livestock are receiving from their ration are not only related to the total amount of energy in the feed but also to the rate of digestion. Forages that are slow to digest will limit the amount of total feed and total energy the livestock can take in. As forages become overmature, they develop more nondigestible fibers such as lignin and thus become slower to digest.

Protein

Protein is an essential nutrient needed for growth, development, and production by livestock. Protein is measured using several different methods and terms. Protein metabolism in ruminants is complex, because protein may be found in different forms depending on whether it comes from digestion of the feed, or instead from microorganisms that pass from the rumen and reticulum where they become food and are digested.

Proteins are made up of chains of amino acids. Ruminants can synthesize some amino acids naturally, but others can't be made by either the ruminant or the digestive microbes in sufficient quantity to meet nutritional needs. These amino acids are called essential amino acids, and the animals must ingest these from their feed. This requires that these essential amino acids must be either in the pasture or in other supplemental feeds provided to the animals.

Rumen microbes can also use nonprotein nitrogen sources, such as urea, to make amino acids. Thus, on a forage test report, the number listed as the overall protein content of a feed represents a combination of many types of amino acids and nitrogen. The overall amount of protein in pasture forage depends on the types of plants growing and their stage of maturity. There is more protein in the leaves of plants than in the stems. Protein content is also higher in less mature vegetative plants than mature ones. Legumes often have higher protein content than grasses.

Lush vegetative pastures can be very high in protein (upper 20s to over 30 percent). Much of the protein is highly soluble, so it breaks down rapidly in the rumen. Increasing the amount of energy in the ration when pasture protein levels are high by supplementing with a feed such as cornmeal can increase the use of these soluble proteins by the microbes and improve the protein-to-energy ratio.

So if supplemental grain is fed along with a high-protein pasture, it should be a high-energy grain. In contrast, supplementing with high-protein grain when livestock are grazing high-protein pasture can create a whole list of problems. For example, when overfed protein, livestock will not graze pastures as well, and may develop hoof problems and metabolic issues. There is more detail on this later in this chapter and in chapters 9 and 11.

Vitamins and Minerals

Livestock require major or macro minerals, trace minerals, and vitamins. Major minerals are calcium, phosphorus, magnesium, potassium, sodium, chlorine, and sulfur. These are necessary for the function and structure of many body tissues and fluids as well as metabolic processes.

Trace minerals needed are cobalt, copper, iodine, iron, zinc, manganese, molybdenum, boron, and selenium. These are needed for hormone and enzyme function and are essential for immune function. Vitamins are needed for immune and metabolic function.

Pasture provides many of these minerals and vitamins, as well as precursors of some vitamins. However, some minerals are usually lacking in the pasture, so most grass farmers provide some supplemental minerals and salt during the grazing season. During the non-grazing season, they may feed a larger amount of mineral mix with higher levels of some important nutrients because lush green pasture has higher levels of some vitamins than stored feeds.

For farms feeding supplemental grain or concentrates, a mineral blend may already be mixed into the purchased grain or concentrate. However, levels of some minerals may still not be sufficient without an additional mineral supplement. It is particularly important for farms feeding low- or zero-grain rations (100 percent grassfed) to provide a well-balanced mineral mix.

If minerals in the ration are not correctly balanced, some health issues can occur. For example, grass tetany is a problem that occurs most often in early-lactation animals that experience magnesium deficiency.

Because soils vary in different regions, it is important to pay attention to what minerals may be deficient in the local forages in your area. For example, in some regions soil selenium levels are very low, so supplementing with selenium can be crucial to avoid health problems such as white muscle disease and higher incidence of retained placentas. In other parts of the country selenium is not deficient, and can even be toxic if fed in levels that are too high. So supplemental mineral feeding should be carefully adjusted to meet the needs of the animals on the specific farm and avoid both mineral deficiencies and toxicities.

WATER

Water often isn't mentioned when discussing nutrients in livestock rations. However, if this essential ingredient in the diet is deficient, it will cause some of the most severe and rapid health problems. The bodies of livestock animals consist of more than 90 percent water, and water is essential to almost every body function. If it isn't provided in sufficient amounts, livestock experience discomfort, will eat less dry matter, and will become sick and die.[1]

Livestock can take in water both from drinking water sources and from pasture plants. Drinking water should be clean and not contain harmful levels of bacteria, nitrates, or heavy metals. Care should be taken when a water tub is situated near an electric fence to ensure that the livestock will not receive a shock when they try to drink!

Pasture plants often contain 85 percent water or higher. Forage tests will report results on both an "as-fed" basis and "dry matter" basis and will report the amount of dry matter versus water in the forage. Because vegetative pasture plants contain so much water, livestock grazing those pastures may consume less drinking water than similar animals being fed stored forages. However, it is always advisable to provide drinking water in or within reasonable walking distance of the pasture.

Testing Forage Quality

Now that we have discussed the nutrients grazing livestock need, let's look at how these nutrients can be measured in forages. With the knowledge of what the herd needs, and what the pasture can provide, you will be able to make well-informed decisions on what supplemental feeds may be needed.

Forage testing can be done of both pastures and supplemental feeds, including forages and grains. Particularly

for a new grazier, it is important to know the nutritional quality of the pasture. This becomes even more essential information when all or most of the feed the livestock gets comes from pasture.

It may be helpful to test the pasture several times throughout the year, particularly when there is a noticeable change in the forage. It may also be useful to test when there are noticeable changes in seasons or weather patterns, or when new forage crops are grown, particularly for alternative crops for which "book values" on nutrition are not readily available.

For a pasture forage test to be useful, it needs to be based on a representative sample of what the livestock will actually eat. Watching them graze is the best way to determine which plants in a pasture the animals are really eating, and therefore what to sample. This way the livestock themselves can provide you with the instruction on how to take samples reflecting their choices of grasses, clovers, and weeds. Avoid collecting from areas contaminated by manure and urine. Also avoid patches of overmature plants, because livestock won't graze those areas.

COLLECTING A SAMPLE

To take a sample that mimics the way livestock graze, grasp some forage, wrap it around your hand, and break it off. Try to achieve the same post-grazing height as left by the cows. This mimics the way cattle wrap their tongues around the plants and then break them off by holding them between their gums and teeth.

Collect 10 to 20 smaller samples in a paddock and mix them together. Then take a subsample from that — enough to pack tightly into the sample bag. Follow the lab's instructions to know what this sample size should be. Many labs will provide you with the sample kit and bag, which is generally about the same size as a quart bag.

Freeze the bagged sample to prevent the plants from changing quality due to continued respiration or fermentation in the bag. After freezing for 12 to 24 hours, send the sample to the lab.

Choose a lab that is certified by the National Forage Testing Association (NFTA). The NFTA maintains a list of certified labs on their website: www.foragetesting.org.

For a perennial pasture mix, most labs will do a basic test for about $20. The Near Infrared Reflectance

Dairy One

FORAGE LABORATORY

DATE SAMPLED	LAB RECEIVED	DATE PRINTED	LAB USE
	09/01/15	09/02/15	.928

ADDITIONAL DESCRIPTIONS
Northeast Pasture

ENERGY TABLE — NRC 2001		
	Mcal/Lb	Mcal/Kg
DE, 1X	1.38	3.04
ME, 1X	1.19	2.63
NEL, 3X	0.69	1.53
NEM, 3X	0.73	1.60
NEG, 3X	0.45	1.00
TDN1X, %	66	

KIND DESCRIPTION	CODE	LAB SAMPLE
Mmg Pasture	012	21909380

DESCRIPTION 1
Northeast Pasture

ANALYSIS RESULTS

COMPONENTS	AS SAMPLED BASIS	DRY MATTER BASIS
% Moisture	78.8	
% Dry Matter	21.2	
% Crude Protein	4.8	22.5
% Available Protein	4.5	21.3
% ADICP	.3	1.2
% Adjusted Crude Protein	4.5	22.5
Soluble Protein % CP		36
Degradable Protein % CP		68
% NDICP	1.5	7.1
% ADF	5.5	25.9
% aNDF	9.5	45.0
% Lignin	1.0	4.7
% NFC	4.1	19.3
% Starch	.9	4.3
% WSC (Water Sol. Carbs.)	2.6	12.2
% ESC (Simple Sugars)	2.5	12.0
% Crude Fat	.9	4.1
% Ash	1.92	9.08
% TDN	14	66
NEL, Mcal/Lb	.14	.68
NEM, Mcal/Lb	.14	.67
NEG, Mcal/Lb	.08	.40
Relative Feed Value		142
% Calcium	.15	.73
% Phosphorus	.08	.40
% Magnesium	.06	.26
% Potassium	.55	2.56
% Sulfur	.06	.29
% Chloride Ion	.12	.58
% Lysine	.19	.88
% Methionine	.07	.31
Horse DE, Mcal/Lb	.23	1.10

Figure 8.1. This pasture forage test report shows the analysis for a sample of good-quality forage. The report on the opposite page is for a sample of a more mature pasture which has a slightly lower quality.

Spectroscopy (NIR) method is often used. In the past, laboratories used a series of chemical procedures to test forages (protein, fiber, minerals). This "wet chemistry" test is still available, but may take longer and cost more. NIR is a technological advance that uses light to more quickly determine the nutritive value of commonly tested forages. However, for alternative or unusual pasture forage tests, it is best to request a wet chemistry test; this will provide more accurate information about the less common pasture plants or mixtures.

UNDERSTANDING FORAGE TEST RESULTS

Forage test report format varies from laboratory to laboratory. See figure 8.1 for examples. A report usually contains information on the following:

- Percent moisture.
- Percent dry matter (DM).
- Percent crude protein (CP).
- Percent acid detergent fiber (ADF).

Dairy One

FORAGE LABORATORY

KIND DESCRIPTION Mmg Pasture	CODE 012	LAB SAMPLE 21909390

DESCRIPTION 1 Ridge Field Pasture		

ANALYSIS RESULTS		
COMPONENTS	AS SAMPLED BASIS	DRY MATTER BASIS
% Moisture	71.8	
% Dry Matter	28.2	
% Crude Protein	5.1	18.0
% Available Protein	4.7	16.8
% ADICP	.3	1.2
% Adjusted Crude Protein	5.1	18.0
Soluble Protein % CP		35
Degradable Protein % CP		65
% NDICP	1.5	5.2
% ADF	9.8	34.6
% aNDF	16.7	59.1
% Lignin	1.1	4.1
% NFC	2.8	10.0
% Starch	.3	1.0
% WSC (Water Sol. Carbs.)	2.3	8.3
% ESC (Simple Sugars)	2.1	7.3
% Crude Fat	.9	3.2
% Ash	2.76	9.78
% TDN	17	60
NEL, Mcal/Lb	.16	.55
NEM, Mcal/Lb	.16	.56
NEG, Mcal/Lb	.09	.30
Relative Feed Value		97
% Calcium	.14	.48
% Phosphorus	.09	.32
% Magnesium	.06	.21
% Potassium	.72	2.55
% Sulfur	.09	.31
% Chloride Ion	.29	1.01
% Lysine	.20	.70
% Methionine	.07	.25
Horse DE, Mcal/Lb	.26	.91

DATE SAMPLED	LAB RECEIVED 09/01/15	DATE PRINTED 09/02/15	LAB USE .934

ADDITIONAL DESCRIPTIONS Ridge Field Pasture			

ENERGY TABLE — NRC 2001		
	Mcal/Lb	Mcal/Kg
DE, 1X	1.25	2.75
ME, 1X	1.06	2.33
NEL, 3X	0.60	1.32
NEM, 3X	0.63	1.39
NEG, 3X	0.37	0.81
TDN1X, %	60	

- Percent neutral detergent fiber (NDF).
- Calcium (Ca).
- Phosphorus (P).
- Percent total digestible nutrients (TDN).
- Net energy calculation for lactation (NEL), expressed as mcal/lb.
- Net energy calculation for maintenance (NEM), expressed as mcal/lb.
- Net energy calculation for gain (NEG), expressed as mcal/lb.
- Relative feed value (RFV).[2]

The information in the test report will generally be presented on both an "as-fed basis" and a "dry matter basis." Most pasture samples contain a lot of water, and that water dilutes the concentration of nutrients. As-fed numbers describe the nutrient concentration of the sample as it was received in the forage lab, including all the water. Dry matter basis figures are those from material after it has been dried. The as-fed values will be lower than the dry matter values, with the exception of moisture. Looking at dry matter basis numbers allows for a more accurate comparison among different forage tests.

Fiber

On forage tests, you will see fiber measured as acid detergent fiber (ADF) and neutral detergent fiber (NDF). ADF is the cellulose and lignin, while the NDF is the cellulose, lignin, and hemicellulose.

ADF (acid detergent fiber) gives an estimate of the energy content in the forage — ruminants obtain a significant amount of their energy from the digestion of these fibers. Lower ADF values mean higher energy and digestibility.

NDF (neutral detergent fiber) predicts how much forage the animals will be able to eat. Forage that is less digestible will fill up their rumen, which limits what else they can eat. Therefore, if the NDF is too high, it will reduce the total amount animals can consume.[3]

For example, a cow can only eat about 0.8 to 1 percent of her body weight in NDF if the forage is lower quality, but she will eat as much as 1.2 percent of her body weight in NDF from higher-quality forage. The percentage she can eat from well-managed, high-quality pasture may be as high as 1.4 percent of body weight in NDF.[4]

Crude Protein (CP)

Crude protein is the measurement most commonly used to see what the feed protein level is. However, it is based on an estimate from the amount of nitrogen in the feed. So it includes both the true protein and the nonprotein nitrogen sources in this calculation. This is calculated by multiplying the amount of total nitrogen found in the forage by 6.25.

Pasture, particularly when lush and vegetative, can be high in protein — often higher than the livestock actually need. This is why it is important to balance the amount of protein with an adequate energy supply, particularly in feeds such as a well-managed pasture with a high legume content.

Energy Terms

It is not easy to measure the energy in forages, so some measures of energy are calculated based on components that can be directly measured in a forage test. In addition to reporting ADF and NDF, forage tests will often include total digestible nutrients (TDN), metabolizable energy (ME), net energy for lactation (NEL), and maintenance (NEM).

NEM and NEL (net energy — maintenance and net energy — lactation) refer to the ability of the forage to meet the energy requirements of different groups of livestock. These parameters are expressed as mega-calories (mcal) per 100 pounds of forage. When balancing rations, dairy producers will probably judge the sample based on its NEL value, while a beef producer would be more likely to use the NEM value.

RFV (relative feed value) can be used to compare forages with one another on an energy basis.

TDN (total digestible nutrients) indicates the percentage of digestible material in the forage.

Minerals

Some forage tests will refer to the total mineral content in the forage as ash. This includes the macro-minerals calcium, phosphorus, potassium, magnesium, sulfur, and salt. It will also include some trace minerals such as iron, iodine, copper, cobalt, manganese, zinc, and selenium.

Interpreting Results

A forage sample taken from the same pasture will be different depending on the time of the year and how mature the plants are. As plants mature, the interior contents of cells decrease in digestibility. At the same time, the cell walls become thicker, adding more lignin (which is not digestible). As the maturation process continues, the overall protein content of the plant will start to decrease.

Thus, forage test results from mature pasture will show high NDF and low protein. A test of lush vegetative pasture will be lower in NDF and higher in protein.

How Much Feed Is Needed?

So far we have been discussing the individual nutrients in livestock feed and overall feed quality. However, it is also important to know what quantity of feed grazing animals need. Knowing the dry matter intake requirement of an individual animal as well as that of the whole flock or herd on a farm helps assure that enough feed from both pasture and stored forages and grain is being fed. In chapter 13, I describe how to use the required amount of dry matter for a herd to calculate paddock size and acreage. This allows a farmer to determine how much land is needed to provide pasture for the farm's whole group of livestock.

Feed intake requirements are described in terms of total dry matter intake (DMI) or dry matter demand (DMD). DMI is the *actual* amount of dry matter an individual animal eats. DMD is the *estimated*, or expected, dry matter intake required for a specific type and class of livestock.

Many sources provide DMD tables and data. This information can be provided in several ways. Information may be listed as a percentage of body weight, as in table 8.1. For example, a dry cow DMD may be predicted to be 2.5 percent of her bodyweight, so a 1,000-pound cow would need 25 pounds of dry matter per day. (DMD = percentage x animal body weight.) Note that this is total required dry matter, some of which may be provided from pasture and some from supplemental stored feed.

In addition to reporting DMD as a percent of body weight, some tables may present the information in pounds per head per day. Also, DMD information for lactating animals may be adjusted based on the amount of milk and quantity of solids in the milk. Table 8.1 is a simplified version to help you get started calculating DMD for your herd or flock, but it is helpful to consult a more detailed table that supplies a more accurate estimate of DMD for your specific type of livestock. Consult appendix E for additional information.

Many factors can change how much an animal will need to eat and how much it will be able to eat. The data in DMD tables can be helpful for making a first estimate of required pounds of dry matter intake per animal or per group, but it is also important to understand and consider factors that cause DMD to vary. In real-life situations, a herd or flock may need to eat more, or less, than these DMD tables predict!

Actual DMD varies depending on the age and growth rate of the animal, how much milk she is making, stage of gestation, breed, environmental conditions (such as temperature), and types (and digestibility) of available feeds. For example, a lactating cow or sheep will eat more than a dry cow or ewe. A cow eating low-NDF forage will eat more than one consuming overmature high-NDF forage.

Table 8.1. Dry Matter Demand (DMD) Requirements

Class of Livestock	Estimated DMD (% of body weight)
Lactating sheep	3.65
Weaned, growing lambs	3.30
Lactating goats	4.00
Weaned, growing kids	2.25–3.30
Lactating beef cattle	2.00–2.50
Growing and finishing beef cattle/steers	2.25–3.35
Lactating dairy cows	2.00–4.00
Dairy heifers	2.50
Dry dairy cows	1.80–2.50

Sources: Lee Rinehart and Ann Baier, *Pasture for Organic Ruminant Livestock*, National Organic Program DMD tables, http://www.ams.usda.gov/sites/default/files/media/Program%20Handbk_TOC.pdf

How Much Are They Eating?

When you're feeding a total mixed ration (TMR) or a controlled ration in the barn, it is easy to figure out how much the animals are actually eating. When some or all of the ration is pasture, however, it is not as easy to determine their feed intake.

One way to determine how much pasture they are eating is to measure the amount of forage dry matter in the pasture before, and then after, a paddock has been grazed. The amount before minus the amount after is the total amount the herd or flock harvested. (Methods to directly measure pasture dry matter are discussed in chapter 13.)

The second method is the feed subtraction method, and it is most commonly used on dairy farms that are providing grain and some forage along with pasture. For this method, begin with the animals' estimated dry matter requirement or DMD. From that, subtract the amount of dry matter actually being fed in the barn from grain and stored forages. This gives an estimated amount the animals will be ingesting from pasture. See appendix E for references on DMI requirements of different types of ruminants.

Grass farmers control how much the animals eat by deciding what, if any, supplemental feeds to provide and by allocating enough pasture. The amount of pasture dry matter available is determined by the farmers' management decisions on how tall the plants are, how palatable the species are, how mature they are, how dense the pasture is, and how large (or small) the paddocks will be.

Supplemental Feeding

Complete details on how to balance a full ration for sheep, beef, goats, and dairy cattle is beyond the scope of this book. However, it's important to address the topic of supplemental feed for grazing livestock, because if it is done incorrectly it will have a negative effect on both the livestock and the pasture. Appendix A has a simple worksheet for troubleshooting some common problems in pastures and livestock, many of which are caused by not correctly providing the right amounts and types of feed to livestock.

If pasture quality is good, and it is managed to provide a consistent amount, then the pasture alone may be able to meet all the forage nutrition needs of many types of livestock. For example, for farms with sheep or beef animals and good-quality pasture, no additional supplemental feed will be needed. On these farms, the ration will consist of pasture, a clean drinking water source, and a supplemental salt and mineral mix. On other farms, however, it may be necessary to feed supplemental forages and grains to grazing livestock, for a variety of reasons.

For dairy farmers, it may be significantly more profitable to feed grain to the lactating animals along with pasture. Feeding zero grain to dairy cows is possible, though doing so successfully requires excellent management, good cattle genetics, and an ample supply of high-quality forage. This is discussed in detail in chapter 11.

Balancing a full ration includes looking at quantities of many nutrients in the feeds, not just total dry matter intake needs. For this reason, farms feeding a more complex ration with multiple types of supplemental feeds may want to hire a nutritionist or learn more about ration balancing. Ration balancing can be done by using a computer program especially designed for the purpose. It can also be done by working through the necessary calculations by hand. In all cases, it's essential to first have the necessary information on the nutritional needs of the animals, as well as on the forage quality of the pasture and other feeds.

In practice, many farmers set up a ration based on their past experience and knowledge instead of using a computer program. It is important to note, however, that a poorly planned supplemental feeding program on pasture can create significant problems for the livestock, the pasture, and farm finances.

Livestock nutritionists can find balancing rations on pasture challenging because it is not easy to determine exactly how much pasture dry matter, and of what quality, the animals are ingesting. Due to the selective grazing habits of livestock, they may be eating what they want from the pasture, not what we plan for them to eat! In addition, the forage quality may vary as the herd or flock is moved from one paddock to the next. If the nutritionist working with a grass farm doesn't take the time to walk out into the pasture to check the quality and quantity of forage available, he or she won't have the information needed to balance a good ration for the herd!

Common mistakes made when supplementing livestock on pasture include inadequate DMI and overfeeding protein. Inadequate DMI can be caused by a variety of issues including pasture quality, pasture height (too overmature/tall or too short), insufficient access to enough pasture per animal, and supplementing with the wrong type or amount of grain or forage.

Overfeeding protein occurs when pastures are high in soluble protein and the protein level in the supplemental feed is too high. This can lead to several problems, including lower pasture DMI, elevated milk urea nitrogen (MUN), loss of body condition, excessively loose manure, and other concerns.

Each farm will need to determine which, if any, feeds to supplement the pasture with. Understanding the needs of the livestock and the quality of the feeds makes it easier to combine the right amounts and types of feeds with the pasture to optimize the pasture quality, profits, and animal performance.

CHAPTER 9

Meeting Livestock Nutritional Needs

Once you understand the nutritional needs of your livestock and know how to test forage and read a forage test report, you can better assess what, if any, supplemental feeds to provide for the animals while they are grazing.

The ultimate goal of many well-managed grass farms is to provide as much as possible of the livestock's nutritional needs from pasture. This requires an understanding of the many strategies we can use to maximize pasture intake, but it is also important to learn how to observe and monitor the livestock to make sure their needs are fully met.

Maximizing Forage Intake

Maximizing the amount of dry matter the herd or flock can consume from pasture is essential to grass farming and farm profitability. For farms that rely on pasture as the primary or only feed source, low pasture dry matter intake (DMI) will lead to poor animal performance, less milk, less meat, less income, and serious animal welfare concerns.

For dairy farms or others who plan to supplement pasture with grain and stored forage, planned DMI from pasture can be lower and still meet production goals. However, it remains important to manage the pasture so that the animals can efficiently harvest it. Every pound of dry matter the animals are able to harvest from pasture is less feed that has to be made or purchased!

In *Grass Productivity*, André Voisin wrote that we should try to "see how the cow proceeds to harvest her food. When we know what method she employs we will be able to help her in her work, that is to say to get a better yield from our pastures."[1] Since grass farmers are asking the cows, sheep, and goats to do the harvesting work, finding ways to make that work as efficient as possible is a priority to maximize pasture DMI.

Maximizing pasture DMI is particularly critical for dairy animals. Because they have high nutritional needs, a shortage of DMI will immediately show up as lower milk production. In contrast, farmers growing lambs or steers on pasture are looking for gains over a whole season, so there is more flexibility in daily DMI and quality. Dry ewes or cows have lower nutritional needs than finishers or lactating animals, allowing even more flexibility in DMI and quality for them.

BIG BITES ARE BETTER

The best way to make sure the herd or flock is eating enough dry matter from pasture is to pay close attention to the quality and size of each bite of pasture they get. More good-quality dry matter per bite will allow the animals to maximize their pasture dry matter as efficiently as possible. Plant height, species, and density are the pasture factors that control how much dry matter the animal can take into her mouth with each bite.

Higher-density pasture, which has more grass tillers per square foot, allows for higher intake levels because the livestock harvest more with each mouthful, and they can find the food easily and quickly. Animals in a high-density pasture can stand in one spot and take bites from many plants with a high biting rate. If there are fewer plants per square foot (lower-density pasture), animals will have a lower biting rate because they are spending more time walking around in search of food. As discussed in chapter 7, a higher biting rate and more dry matter per bite adds up to more DMI.

Increasing the density of the pasture by managing to encourage more diversity, higher plant density and grass tillers per plant was discussed in chapters 3 and 5. More tillers allows for more leaves, which capture sunlight and fuel healthy plant growth. But from the animals' perspective, more tillers and plants is also an essential part of making it possible to most efficiently harvest pasture dry matter. Once again we see how good management of grazing-adapted plants is equally beneficial to the plants and the grazing livestock.

Plant density may be low in pastures that have been recently planted or converted from hay crops to pasture. Low density results in reduced amounts of dry matter per bite and a decrease in total DMI for the animals. In a very low-density pasture, it may not be possible for some animals to meet their dry matter intake needs even if they are given a larger area in which to graze. Time will become the limiting factor, because each bite harvests such a small amount that the animals can't fill their rumens.

In addition to pasture density, pasture height has a large influence on pasture DMI. Grazing livestock do not consume plants from tip to ground level. Dairy cows usually harvest only the top third of pasture plants. If the pasture is too short, when a cow bites off that upper part of a plant, she may take in only a small amount of dry matter. For example, cows turned into a 3-inch-tall

pasture may remove only 1 inch of plant material per bite. This is stressful for the plants, which most likely have not fully recovered from the previous grazing. It is also not good for the cows, which have to expend energy for a meager return of tiny amounts of forage per bite in the quest to fill their rumens. It is important to make each bite count!

When animals are left in the same pasture for several days or a week, the nutritional quality of what they eat each day changes due to selective grazing behavior. On day one, they will find very high-quality feed, which takes them little effort to eat. Several days later in the same pasture, the cattle will be working harder but getting lower-quality feed. Using a higher stocking density (smaller paddocks) and moving the animals to new pastures more often will result in more predictable pasture nutrient intake, which can make ration balancing easier and milk production more predictable.

If pasture quality and digestibility are high, then pasture intake will be high. If the pasture is overmature and contains too many fibrous stems and not enough leaves, then it may not be digestible enough. This low digestibility can also limit the animals' ability to consume enough pasture dry matter.

Feeding too much of the wrong types of supplemental feeds in the barn can also reduce pasture dry matter

Figure 9.1. Livestock in this pasture are able to rapidly fill their rumens with high-quality forage. The animals can maximize dry matter per bite because the diversity, density, and height of these plants are ideal for grazing efficiency.

Figure 9.2. This pasture has been damaged by years of overgrazing, so these plants won't grow much taller than this even with an extended rest period. With the plants this short, dairy cows don't get much dry matter per bite!

intake. Dairy farmers new to grazing will often feed too much protein or forage in the barn, and will then be discouraged that the cows don't want to eat much pasture. Overfeeding protein not only causes health problems for the animals but also decreases their desire to graze the lush high-protein pastures, and may cause their DMI to be too low.

WASTE CAN BE BENEFICIAL

If the goal is to maximize pasture DMI, particularly with finishing growing animals or grazing cows that are making a lot of milk, it is much better to move animals frequently and leave some percentage of palatable pasture material in each paddock ungrazed. Don't push the herd to graze it down too short. Leaving more plant residual has the effect of ensuring quality feed for the livestock — they get the best from each paddock in which they graze. This practice of leaving more plant residual intact is also better for the plants because it ensures they are never grazed down too short.

In terms of "wasted" pasture, livestock will always leave some ungrazed plants around areas of manure. Also, given the choice they won't graze plants all the way down to the soil surface. Neither of these is necessarily a bad thing! These natural instincts help livestock avoid eating the less digestible lower portions of the pasture

and plant material that might contain parasites. The best way to manage rejected areas around manure piles is to improve the biological activity of the soil and population of insects such as dung beetles so that manure is more rapidly incorporated into the soil. Using a higher stocking density so that the leftover plant residual is trampled will encourage this decomposition activity.

For farmers who can't use a high stocking density to achieve a thorough trampling effect, clipping pastures immediately after grazing may help manage rejected forage and standing residual, particularly in the first few years of grazing. However, clipping must be done *immediately* after grazing, *before* the plants begin to regrow. If plants are clipped after regrowth has started, they will be damaged. It is also important *not* to clip the plants too short. Resist the urge to mow pastures so they look like a well-kept lawn. Pastures that look a bit messy are often very healthy and productive! Most pasture plants will do best if they are not clipped any shorter than 3 or 4 inches.

Clipping with a mower can be a useful tool for both weed control and mowing straw-like overmature grass stems so that regrowth at the next grazing is higher quality. That next grazing will then consist of more palatable, digestible leaf material, which can allow the animals to fill up faster.

Figure 9.3. This pasture has just been grazed by the dairy herd, leaving behind enough plant material that the plants haven't been damaged, and the cows didn't have to work too hard to fill their rumens.

Figure 9.4. On this farm, most paddocks are clipped once a year to control weed species. Clipping is done immediately after moving the herd into the next paddock to avoid clipping any plants that start regrowing quickly.

Monitoring Livestock Well-Being

So far we have discussed the basic nutritional needs of livestock, how to know what nutrients are in their feed, and how much dry matter they require. We have also talked about some methods to increase the amount of feed they get from the pasture. But how do we know if we are successful at meeting the herd's or flock's nutritional needs?

Grass farmers need to learn how to monitor and observe the well-being of the livestock to make sure the management and feeding systems are keeping animals healthy and productive. The sooner a problem can be observed, the easier it is to correct it. And when it comes to the production of meat and milk from grassfed animals, not catching a problem early can be expensive and potentially lead to animal welfare issues.

Several monitoring and testing tools can be useful for assessing livestock to see if they are being fed correctly. These include visually assessing rumen fill to make sure livestock are getting enough total dry matter, looking at the manure to see how digestion is functioning, looking at forage tests of the pasture to see what the feed quality is, body condition scoring (BCS), watching herd or flock behavior, measuring milk production per cow or rate of gain in growing animals, and monitoring milk urea nitrogen levels. Appendix A in this book includes a troubleshooting worksheet to help think through some of the most common pasture-related problems, and appendix C is a worksheet on monitoring to see if the pastures themselves are getting better over time.

Direct measurements of how much dry matter is actually available in the pasture when livestock are put in the paddock, and then measuring how much is left behind, can help determine how much total dry matter the herd or flock is harvesting. Direct measurement of pasture dry matter is discussed in chapter 13.

Observing Rumen Fill

As discussed in chapter 7, it is possible to see how full an animal's rumen is by looking at the animal's left side. To start with, observe the left side to see whether it looks round and full or less distended and a bit "empty." You can also get a good sense of how full or empty an animal is simply by standing in front of or behind her and observing her overall diameter or how round she is.

Also, look just under the short ribs at the triangle of soft tissue on the left side of the animal. This is where the rumen is not covered up by ribs, so you can see if it is sunken and empty, nicely filled up, or perhaps overinflated (see figure 9.5). An overinflated rumen can indicate a serious issue with bloat, which requires immediate medical treatment. A sunken or empty-looking rumen while an animal is on pasture can mean that the

Figure 9.5. Look at the triangle of soft tissue under the short ribs of the brown cow in the left-hand photo. It is not as full as the same area of the cow in the right-hand photo.

animal is not getting enough feed due to forage availability, quality, or illness.

Body Condition Scoring

Body condition scoring (BCS) is an important management tool that can help you evaluate health, optimize production, and determine the nutritional status of an animal on pasture. BCS is based on the amount of body reserves, or fat and muscle, that an animal has.

Body condition is determined by a visual and hands-on examination, and is quite accurate when done by a trained evaluator. Farmers can be trained in this useful technique and become proficient at it on their own farms. A visual exam can provide some information on body condition, but particularly for animals that have long hair or wool, hands-on evaluation is needed to assure accuracy.

The end result of BCS is a numeric score, which can be used in several ways. First, it can be helpful to track changes over time in animals as they go through the grazing season. It can also be compared with the standard desired score for an individual animal's stage of production. This can help with decisions on how much supplemental feed to provide and how much pasture dry matter to give to the herd, and with other important livestock and pasture planning decisions.

Body condition scoring is different for each type of animal. However, the approach to doing the scoring is very similar regardless of the species. For example, beef cattle are evaluated using a scoring a system from 1 to 9, while the system for sheep and dairy cows uses a range of 1 to 5. On both scoring systems, the lower numbers correlate with thinner animals.

Body condition will naturally vary throughout an animal's production cycle. Dairy cows at peak lactation, for example, will be thinner than when in their dry period. Other factors that influence body condition include weather, housing type, feed intake, feed quality, and illness.

Regular scoring of the herd or flock can allow farmers to catch problems before they become serious. Once an animal drops a full condition score, it can take a lot of time and calories to get back into condition. If she stays too thin for too long she may not breed back well or make

much milk, which will affect farm cash flow. See appendix E for additional sources of information about BCS.

Manure Scoring

Observing manure can provide valuable information on how the nutritional needs of the livestock are being met, how well balanced the ration is, and what the nutritional levels in the pasture are. Manure scoring is particularly helpful in showing us if too much protein is being fed, and how well the fiber or supplemental grain in the ration is being digested. For example, pasture is high in water (80 percent or higher), low in dry matter, and when vegetative is high in protein and low in fiber. Livestock feeding on vegetative pasture will frequently have loose manure. However, if the flock or herd is grazing overmature pasture, the manure may become very stiff and be full of undigested fibers.

Manure scoring can only be done for weaned animals because manure texture is different while they are drinking milk. Manure scoring is done on a scale from 1 to 5. The scale starts with 1, which is the very liquid manure from cattle on lush spring pasture, or cattle with diarrhea. Manure with a score of 5 is very stiff and firm. See figures 9.6 through 9.10. These photos and descriptions are for cow manure; a somewhat similar scoring system can be used for goat and sheep manure (figures 9.11 through 9.13), which has a different texture than cow manure. Any of these scoring techniques are easier to do if you are willing to touch the manure and think of it as an interesting and useful farm resource, instead of something gross!

Score 1 is manure that is so liquid it may actually arc as it passes out of the cow. The manure forms a wet puddle or splatters widely in the pasture. If you are willing to put your fingers in the pie, it will feel creamy and smooth without any undigested fiber.

Excess protein in the ration or lack of fiber can result in this type of manure. This can commonly occur on lush spring pasture, and if a high-protein supplemental feed is also being used it can be made much worse. Feeding some forage or a higher-energy grain can make the manure less runny. Animals with diarrhea caused by an illness may also have manure of this consistency, so if it continues even as the pastures mature or the ration is changed, it may be necessary to consult with a veterinarian.

Score 2 manure is still runny, won't form a nice pile, and will splatter as it lands in the pasture. If you put your finger in this manure, it will also feel creamy and smooth but it may have some particles of undigested forage in it.

This manure is what is likely to be seen from cows on pasture once the spring flush of very lush vegetative pasture is done. Manure will be less runny if they are getting supplemental forage along with the pasture.

Score 3 manure will form a nice pile that has rings around a small depression in the middle. This manure is somewhat like a pudding. It doesn't feel creamy and smooth like score 1 manure, and it contains undigested fibers from the forage.

This is the manure that makes plopping sounds as it lands in the pasture, and it will stick to you and the cows if it lands on you. This is the optimal score for a cow pie.

Figure 9.6. Manure score 1 is runny like pea soup and often splatters widely in the pasture.

Figure 9.7. Score 2 is runny and won't form a nice pile.

Figure 9.8. Score 3 manure forms a perfectly round pile with a small depression in the center.

Figure 9.9. Score 4 is thicker and stacks up taller.

Score 4 manure forms a thicker, taller pile than score 3 manure. The pie will not have a depression in the center or concentric rings. Undigested food particles and fiber are clearly visible.

This type of manure may indicate that the diet is low in protein, and will more likely be produced by dry cows, older heifers, or other groups being fed lower-quality pasture.

Score 5 manure is firm and stiff and may even form a ball. This manure contains larger undigested food particles, and you may even be able to see large identifiable pieces of fiber or grain from the ration. Animals eating very poor-quality forage produce score 5 manure, but it can also indicate a blockage of the digestive system or dehydration.

Sheep and goats tend to have firmer manure overall, which will often be in pellets. However, there will be

Figure 9.10. This score 5 cow manure has been rained on, which makes the undigested fiber and food particles more visible.

Figure 9.11. This sheep manure is from a flock of dairy sheep eating lush pasture, and is an example of the sheep equivalent to score 2 manure.

Figure 9.12. This sheep manure is from animals grazing a pasture with slightly lower protein levels and more fiber, so the manure is similar to a cow manure score of 3.

Figure 9.13. Goats and sheep will often have pelleted manure when grazing a balanced diet of forages.

similar textural changes in manure as the diet changes. Sheep and goat manure will range from firm dry pellets to soft clumps. If the manure of the sheep or goat flock becomes runny or progresses to diarrhea, illness such as internal parasites may be a problem, and those animals should be tested immediately.

LIVESTOCK BEHAVIOR

Watching how the herd or flock behaves on pasture also provides useful information on animal well-being. Understanding grazing behavior can also help improve the grazing management system. Grazing behavior is discussed in detail in chapter 10.

Well-fed, healthy livestock on pasture should be energetic, bright-eyed, and socially interactive with one another. Animals that keep to themselves, or do not move into a new paddock with curiosity and enthusiasm for fresh new grass, should be carefully watched. Cows, sheep, and goats are herd animals; when one animal starts separating itself from the rest of the group it may be ill, about to give birth, or suffering from a nutritional problem.

Group behavior can also be significant. If the herd or flock are at the gate yelling when you come to move them to a new paddock, it may be a sign that there was too little dry matter for them in their current pasture area. Or they may be complaining about the forage quality. On the other hand, if they are obviously full and quickly lie down to chew their cuds as soon as they move into the next paddock, it can indicate that the paddock may have been too big.

Cud chewing is an important behavior to watch. Cows on a high-forage diet will spend about a third of their time chewing their cuds.[2] Regularly assessing how many animals are chewing their cuds while resting provides helpful info on how their health is, and how good their diet is. If few animals are observed ruminating (chewing their cuds), then they may not be getting enough forage or fiber levels may be too low. They may also be avoiding eating due to illness.

GROWTH RATE AND REPRODUCTION

Reproductive performance and growth rates are another measure of how well animals are doing. For growing animals such as lambs or finishing steers, watching to make sure they are gaining weight at the right speed is one of the easiest ways to make sure they are having their nutritional needs met. Growth can be monitored by visual observation, which requires less labor than actually weighing the animals.

However, in order to really be accurate, running the group through a scale a few times during the grazing season will provide much more accurate information on rate of gain. The ideal way to do this on a grazing operation is to have a portable scale and livestock corral. This can be brought to the pasture where the herd or flock is grazing to minimize livestock stress and unnecessary walking.

In mature animals of breeding age, keeping track of how quickly they breed back after birthing provides useful information on how good their health and nutrition is. Particularly for a dairy farm, monitoring days in milk (how long the lactation is), calving interval, and the number of open cows is important.

MILK PRODUCTION AND COMPONENTS

For dairy farmers, monitoring daily milk production is one of the easiest ways to make sure animal nutrition and health are good. This can be done for the herd as a whole by checking the amount of milk in the bulk tank; reviewing individual milk weights is also effective. If the amount of milk is fluctuating from day to day, the total amount of pasture and quality may not be consistent. For example, if a farm gives the herd a fresh paddock every three days, milk weights may be highest on day one, and lowest on day three. This would be a clear indication that the paddock is too small or that the grazing system should be adjusted to provide fresh pasture more frequently.

Knowing how much milk each cow in the herd is making provides very useful information and can help assess individual animal performance and allow for grain or other supplemental feeds to be adjusted based on production. Some milking systems provide individual weight information. However, it is often necessary to hire a milk testing service to come to the farm to get individual milk data on each cow. Another option is to rent equipment from a milk testing service and do the testing regularly yourself.

Milk testing services can also provide other useful information on milk solids, somatic cell count (SCC), and milk urea nitrogen (MUN). Some milk buyers also provide all this data on milk components to the farm regularly.

Factors that influence the milk components include genetics, stage of lactation, and what is being fed. Butterfat is strongly influenced by what is fed, because the fats are partially created by digestion of fiber by rumen microbes. Thus the amounts and types of feeds result in a change in the amounts and types of fats made. When the rumen microbes have plenty of fiber, they produce acetic acid, which is used in the udder to make fat. A high-forage diet should result in higher butterfat. Milk protein is also related to diet, but is more strongly influenced by genetics.

When cows eat too much protein and are not getting enough energy in the ration, rumen microbes make ammonia; this passes from the rumen, and some of it ends up as urea in the milk. This milk urea nitrogen can go up if the cows are either eating too much protein or not getting enough carbohydrate energy.

MUN levels can be a good troubleshooting tool, although there is some disagreement about what ideal MUN levels should be. In general most nutritionists recommend MUN of 8 to 12 mg/dl. On pasture, particularly on lush spring growth, it can be very difficult to prevent MUN levels from elevating into the upper teens. Animals on lush pasture that are receiving a high-protein grain can have very high MUN, which is a warning of potentially serious health problems. On the other hand, if MUN levels are very low, the feed may be too low in protein.

The Art of Good Grazing

Making Butter from Pastured Cows

Kriemhild Dairy Farm in central New York milks almost 400 cows. During the grazing season, the farm provides most of the herd's feed from pasture. The milk from the herd is used to make butter, and with increasing consumer interest in the human health benefits of grassfed butter, demand is steadily increasing.

The cows are managed in two groups during the grazing season. A group of older cows graze near the barn, while the younger cows get to walk a little farther to reach pasture, and this includes walking through an underpass, as shown in figure 9.14. The view of several herds of multicolored cows out on the lush green pastures sometimes slows traffic on the road that runs through the farm. But thanks to the underpass, most passersby have no idea that hundreds of those cows walk right under the road several times each day.

Once through the underpass, the young-cow group has to walk uphill through a series of lanes to get to the pasture that they will graze. The older cows graze in pastures that are closer to the barn and less steep so they don't have to walk as far or hike up hills. Both groups are moved to fresh pasture after each milking, twice a day. Some of the pastures on the farm are a full mile from the barn, so these cows get quite a bit of exercise along with their high-quality pasture forage.

Figure 9.14. This underpass was a significant investment for Kriemhild Dairy Farm, but it's essential so that both sides of the road can be grazed. Photo courtesy of Fiona Harrar, Kriemhild Dairy.

Figure 9.15. The dairy pastures include a diverse mix of several species of cool-season grasses, legumes, and forbs.

Figure 9.16. Plant density is very high in most pastures, as is seen in this photo of a pasture that has just been grazed by the herd.

During milking the cows also consume a small amount of barley (up to 7 pounds per cow per day) with minerals. They also get a little balage in the barnyard while they wait to go back to pasture. Each pasture has at least one water tub, but in many areas the farmers try to provide access to two large tubs due to the herd size. They also use 1¼-inch water pipe so that the flow rate is fast enough to keep up with demand.

Farm owners Bruce and Nancy Rivington set up the farm so that the herd is seasonal; the entire herd calves in late winter and early spring. This allows the herd's peak nutritional needs to be timed for when the pasture quality and quantity are highest. Lush high-protein pasture provides the majority of the herd's dry matter intake needs during the grazing season, with the barley providing additional energy. In early lactation, the herd is milked three times a day. Once the grazing season gets seriously under way, however, they switch to twice-a-day milking.

The farmers have selected for cows that produce high-butterfat milk on a high-forage diet. The herd includes Ayrshires, Canadiennes, Dutch Belts, Linebacks, Jerseys, and some Holsteins. They have a focus on calm animal handling and low stress, and Bruce spoke highly of what he learned at a livestock handling workshop taught by Bud Williams.

The farm's pastures consist entirely of cool-season perennial grasses and legumes with some forbs mixed in. Each area is grazed for no longer than a day or two, and the farmers make use of polywire so they can efficiently provide fresh pasture after each milking. Pastures are left to regrow for about two weeks in the spring when plant growth rates are fastest. Later in the summer pastures are allowed to regrow longer between grazings, and some of the land that was hayed earlier in the season is added into the grazing rotation. Pastures are managed so that they are grazed when they are vegetative and leafy. The farmers try not to let the cool-season perennial grasses get too tall and go to seed in the spring. Bruce said he aims for a pregrazing height of no taller than 8 to 10 inches, so the longest regrowth period on the farm is about 35 days. He found that the cows do best at this height. They seem to get less dry matter if it is much shorter than this, but the pasture plant density was lower when he tried taller pregrazing heights. Given the types of plants in the pastures on the farm, Bruce found that this is the ideal pregrazing height.

When Bruce and Nancy started farming in New York 15 years ago, they planted perennial ryegrass on the farm since it is such a high-quality cool-season pasture forage. The farm had been previously growing corn and alfalfa in rotation, so it took some investment in seed and labor to get it revegetated. Their goal is to have a diverse mix of plants with a focus on high plant density. The pastures now include a mixture of Kentucky bluegrass, perennial ryegrass, white clover, and several forb species including dandelion and plantain.

The Rivingtons and their farm staff keep records of where they graze, and for how long. This helps them make sure they are not overgrazing any of the pastures and gives them information for next season's grazing planning.

CHAPTER 10

Grazing Behavior

Most people who watch a group of cows or sheep on pasture probably think that the animals are just randomly wandering, eating, drinking, and taking naps. However, what is actually happening is an incredibly complex relationship between the grazing livestock and the plants in the pastures.

Grazing-adapted plants and grazing animals (herbivores) have coexisted and coevolved for millions of years. But this relationship isn't always friendly! As discussed earlier in this book, livestock grazing can damage or kill plants when not managed well. Similarly, grazing livestock can eat too many plant toxins and become ill or die.

By gaining a better understanding of foraging behavior, the characteristics of plants that influence whether animals eat them or avoid them, and the factors involved in how herbivores choose what to eat, grass farmers can improve grazing efficiency, profitability, and animal welfare on the farm.

Before taking a bite of a pasture plant, an animal investigates it to determine whether it feels palatable, tastes good, smells good, and is nutritious. The animal uses nose and whiskers to start with, and then the tongue. You can watch this when a herd or flock has just moved into a new pasture. The animals use their noses to feel stems, stubble, and other tough material, which they will try to avoid. They will also avoid areas where there is fresh or decomposing manure, and they will probably eat very little of plants that they are not familiar with. They will pause to graze when they find soft leaves, which are more palatable and can easily be torn off the plant.

Unlike ruminants being fed a total mixed ration (TMR) in a feedlot or confinement barn, grazing

PEACEFUL COEXISTENCE

The flock at the Flack Family Farm has grazed for many generations in the pastures there, which include poisonous plants such as false hellebore. It is a closed flock, so all the sheep were born on the farm and grew up with adult sheep that already knew which plants they could eat. The flock is well adapted to the farm climate and plant species. In addition the pastures on this farm have been very well managed, so there is a steady supply of high-quality forage throughout the grazing season. This assures that there are always safe and palatable pasture species for the sheep to eat so they are not tempted by hunger to try the poisonous plants. As a result of this careful management of both sheep and pastures, there have been no mortalities from false hellebore.

Figure 10.1. This sheep flock is grazing amid false hellebore, a toxic plant.

livestock have many choices about what, when, and where to eat. In a pasture or rangeland setting, they encounter new plants, which may be nutritious or nondigestible, potentially toxic or safe to eat as long as they don't eat too much. The way they make choices about what to eat may seem simple to the casual observer watching cows or goats in a pasture. However, their decisions on what to eat and what to avoid are based on a complicated and amazing process that is anything but simple.

An animal that makes the choice to eat a new plant it has never seen before may get sick, may die, or may find that it has found a new nutritious food. Each time the herd or flock is moved to a new pasture, they may encounter new plants. In order to survive, they need curiosity to try new things so they don't go hungry. But they must also be able to learn and adapt, so they don't eat too many toxic plants. In addition, the nutritional content, digestibility, and toxicity of plants changes during the grazing season, sometimes rapidly. So even when coming back to a pasture they have been in before, they may be encountering very different forages.

So how do they learn their grazing behavior? Young animals learn some behaviors as they watch what Mom eats. But what happens later when those youngsters encounter a new plant and Mom isn't there to help them learn if it is safe to eat? If they eat something new, then they will learn from the consequences, or how they feel after grazing something new (post-ingestive feedback). But animal behavior is complex, so there are a lot of other variables involved. We also have to think about animals' genetic predispositions, and where a particular group of animals came from or evolved. Animals' nutritional needs change depending on how old they are and if they are growing, lactating, or gestating. They also change if they are ill or infected with a high level of parasites. Animals' nutritional needs also change depending on age and stage of life, which will influence what and how much they eat. Animals get tired of eating the same plants all day every day, and will enjoy some diversity and variety.

Finding Nutrients, Avoiding Toxins

The nutrient supply in a pasture is constantly changing as plants mature and shift from producing leaves (easier to digest) to producing more stem material (harder to digest). Some plants may also produce toxins, and some species can produce them in high levels, which can be acutely poisonous to livestock. However, many other plants have low levels and can be eaten either at certain times of the year, or as part of a diverse forage diet. The amount of toxins will vary depending on their stage of growth, time of year, and weather conditions.

Livestock have to adjust to these constantly changing levels of nutrients and toxins, even as their nutritional needs are constantly shifting for the many reasons mentioned in chapter 9. As managers, grass farmers can make the animals' job easier or harder through their choices of where and when they move the herd or flock, what plant species they plant or encourage, and how they manage pasture rotations. More diversity in the pasture gives the animals more to choose from, and makes it more likely they can find enough variety to eat without consuming too many toxins. Making management choices that supply the livestock with more high-quality pasture keeps them well fed, which decreases the chance they will try to eat something poisonous due to feeling hungry.

POST-INGESTIVE FEEDBACK

Palatable foods taste, feel, and smell good. Because they are pleasant or at least acceptable, they are foods that animals choose to eat. To an animal in a pasture, plants that are vegetative, thornless, and soft enough to bite easily are palatable. But there is more to flavor and palatability than just odor, texture, and taste. The experience of "good" flavor is also related to feedback the animal has received about the nutrients and toxins in the plant. This post-ingestive effect of plants on the animals plays a large role in how livestock "decide" which plants to eat when and in what quantity. An animal's past experiences with the food, its current nutritional needs, and the nutritional and chemical properties of the plant all play a role in this feedback process. This process, in combination with the animal's ability to use its senses of smell, taste, touch, and sight, helps in the decision-making process.

Sometimes the consequence of eating a plant is positive because an animal feels good after eating. Its nutritional needs were met by eating some plants; it had a positive experience. This positive reinforcement, or post-ingestive satisfaction, increases the chances this animal will want to eat these plants or similar-tasting plants again in the future.

But if the animal eats a plant that contains toxins and becomes ill and feels discomfort in its rumen, this is negative reinforcement. The next time it's in a pasture with those plants, or even plants that have a similar flavor, it may choose not to eat them, or may eat less of them. The extent of change in this animal's behavior will depend on how bad it felt and how many times it ate the plant and felt sick after. It is also more likely that the animal will associate the plant with feeling sick if it felt bad fairly soon after eating the plant.

Many plants contain some toxins at least some of the time. Animals are able to eat a certain amount of them without getting sick. And even valuable nutrients can be toxic if an animal ingests too much of a particular chemical. For example, too much high-protein feed can cause illness if not balanced with an energy source. But when grazing in a diverse pasture of plants they are familiar with, animals will naturally choose to eat a variety of plants so they don't consume too much of any single nutrient or overingest plant toxins.

TOO MUCH OF THE SAME FOOD

There is safety in eating the same familiar foods, but we all get sick of the same food when we have to eat it all day, every day. So once satiated on a familiar food, eating something new or at least different is a common behavior. This is one of the reasons that animals in a pasture with diverse plant species will do better than those in a monoculture.

Behaviors relating to food choices are also strongly influenced by what else the animals have had to eat recently. Animals fed a high-energy forage or grain in the barn will go out to pasture and seek out high-protein plants. If fed high protein in the barn, they will instead search for high-energy plants. This is just one example of how "taste" or palatability is influenced by factors within the animals as well as plants.

PLANT DIVERSITY PREVENTS PROBLEMS

Increasing the number of different types of grasses, legumes, and forbs in the pasture makes it easier for livestock to balance their nutrient needs and avoid eating too many toxins or too much of one nutrient. Diversity also can encourage increased forage intake. Fred Provenza wrote extensively on this subject in *Foraging Behavior*:

Biochemical diversity adds spice to life for livestock, improves economic viability for producers, and maintains the ecological integrity of agricultural landscapes. To meet nutritional requirements, animals need a variety of foods. The kinds and mixtures of plant species influence food intake and animal performance. Offering animals a variety of foods on pastures and rangelands helps each individual to meet its nutritional needs. Individual herbivores, when given a variety of foods, balance the ratio of macronutrients in their diet to meet their nutritional needs. Turnips in ryegrass pastures and grass-legume mixtures can help livestock maintain a better ratio of energy to protein while minimizing effects of toxic compounds in plants. Providing a variety of foods that differ in macronutrients also allows for changes in nutritional needs, such as changing demands for milk production and daily variation in activity and weather.

When foods contain different kinds of toxins that are complementary — that is they operate on the body and are detoxified in different ways — they may have a positive influence on food intake and animal performance. Forages like white clover contain cyanogenic compounds that limit intake by herbivores. Endophyte-infected tall fescue produces alkaloids that adversely affect food intake and livestock performance. Cattle in Missouri performed better on fescue and clover pastures than on legume-only pastures because the mixture contains complementary toxins. It may be beneficial to plant forbs like sainfoin that contain tannins together with legumes like alfalfa that cause bloat. That's because tannins and proteins that cause bloat form stable complexes in the intestinal tract, thereby reducing the amount of foams that cause bloat. We have much to learn about how animals might mix their diets to reduce toxicosis. We also have much to learn about biochemical complementarity among plants in mixtures and how concentrates fed in confinement affect selection of forages in pasture. No doubt our lack of knowledge contributes to observations that a plant is palatable under some conditions and unpalatable under others. Palatability depends on biochemical interactions among the mix of foods available.[1]

Plants That Poke or Stab

Some plants have mechanical defense mechanisms such as thorns that make them less palatable, and protect the plant by discouraging grazing or browsing. Goats and sheep are better adapted than cows to graze on thorny plants because they can carefully select the more palatable individual leaves from stems while avoiding the nonpalatable thorns.

Thorns are obvious plant parts that discourage animals from eating them, but plants can also defend themselves by being hairy, spiny, rough, or even irritating to the tissues of the animals' mouths. You can test for mechanical defense mechanisms by taking a leaf from each grass species in a pasture and chewing a bit on each one in turn. Some will be sweet and soft; others will be hairy or coarse or even a bit painful to chew. Perform such tests carefully if you are not familiar with which plants may be toxic. However, once you know which plants are present in the pasture, and which are common nontoxic forage species, you can discover a lot of useful information by chewing on what your herbivores have available on their grazing menu. For example, it may suddenly make sense why the animals don't seem to eat the overmature orchard grass in a pasture — even though it is not toxic, the texture and flavor of the leaves isn't very palatable!

Other plant parts that can slow or decrease forage intake include standing grass stems or stubble that pokes livestock in the nose as they try to eat. This is one of the reasons farmers managing high-performance livestock such as dairy cows often clip nontrampled post-grazing residual. Clipping can eliminate coarse stems and stubble, which might slow intake rates the next time the herd grazes the paddock.

Learning from Others

Animals learn a great deal of grazing behavior from the rest of the herd or flock. Dr. Frederick Provenza wrote that "a young animal's interactions with its mother and peers have a lifelong influence on where it goes and what it eats."[2] A cow that spends its young life in a barn or feedlot being fed only dry hay and grain will not immediately recognize pasture as a source of feed if it is moved to a new farm and turned out into a paddock.

Anyone who has moved a group of animals from a confinement feeding system out onto pasture for the first time has already learned this lesson. Animals that have never been on pasture will walk around in lush, green, nutritious pasture but won't graze, because they don't know what to eat. They may moo and bleat and get hungry, and if they're not carefully managed they may get sick from lack of feed during the adjustment process. If that same group of animals is moved into a pasture that contains some toxic plants, they are also more likely than pasture-savvy animals would be to eat those plants and become ill, because they have no previous experience with them. Mom didn't teach them about these plants, so not only do they not know which ones are good to eat, they may not know that some are toxic. Animals that don't cautiously approach new plants may quickly end up poisoned and dead. On the other hand, if they are too cautious about new plants, they may become ill or die from lack of nutrition.

These behaviors can also be more subtle. Goats that grow up eating browse species on a farm in one location may not thrive on another nearby farm, where the woody browse species appear to us to be similar. Despite visual similarities, if the pastures on the two farms contain different species that these goats have never encountered, they will not immediately start eating the new kinds of plants and may instead get hungry, lose some weight, or escape through the fence and into the neighbor's garden. Animals in a pasture with unfamiliar plants will also spend a lot more time walking around looking for something to eat. They will have a lower biting rate and lower intake rate.

Behavior During Transitions

Once livestock become familiar with a farm and its pasture rotation, their stress levels will drop, their pasture intake levels will increase, and they will be less likely to eat too many toxic plants and get sick. They will learn which plants are good to eat at different times of the year, and which ones are unsafe to eat. Then the next generation of lambs, kids, or calves born on the farm will learn that same information from their mothers and probably perform even better.

The real challenge arises when moving a herd or flock to a totally new farm or to a pasture full of plants they have never encountered before. This is stressful to the animals in many ways. They are in a totally new environment and may be mixing with new, strange livestock. They probably also experienced stress from riding in a cattle trailer, and now they have to figure out where to find shelter, water, and feed.

Several strategies can ease the stress of transition and allow the animals to adapt more quickly to the new environment and types of plants. Some of these strategies are basic good animal husbandry, including low-stress handling and working with a vet to make sure the animals are healthy before the move happens. Introducing some of the new feed the animals will encounter before they arrive at the new farm may help them adjust. It can also help to bring some familiar feed with the animals on their move. This will give them something they can eat without hesitation as they learn the new foods.

If the move is a significant change for the herd or flock, they will need at least a couple of weeks to adjust. Some farmers report that animals take a full year to adjust because they have to learn the foods and routines

of all the seasons on the new farm. And some animals just won't adjust and may need to be culled.

Helping the Herd or Flock Thrive

All this information on why the cows, sheep, and goats are eating or not eating various plants and food is useful, even though some of it is pretty complicated. So let's summarize how grass farmers can use this information on herbivore behavior to provide balanced nutrition to the herd or flock, increase overall pasture DMI, minimize intake of toxins, and reduce animal stress.

- Offer a diversity of plants, including forbs, grasses, and legumes.
- Let the young stock graze with their moms and other animals so they can learn about the forages on the farm.
- Don't overfeed protein in supplemental feeds when also providing the animals with lush high-protein pasture.
- When bringing new animals to the farm, do everything possible to minimize stress, and if possible provide them with familiar forages as they transition to the new environment and new pasture species.
- Use a "training yard" to train young or new animals to learn about electric fence safely before they go out to pasture.
- Make sure there is high voltage on the electric fence, and that it is well grounded. This is most important when training new or young animals to the fence. But the behavior of the herd or flock will be much more compatible with short-duration grazing (with frequent moves to new pasture) if they are well trained to respect the fence.

Appendix E includes sources that provide more information on how to apply herbivore behavioral knowledge.

The Art of Good Grazing

Micro-Dairy Grazing System

Earthwise Farm and Forest, located in central Vermont, is a certified organic diversified family farm and forestry operation. Carl Russell and Lisa McCrory, along with family members, use draft animal power to grow vegetables, herbs, and forest products. They also raise and sell poultry, pork, beef, and raw milk. Lisa, who is also a grazing consultant, teaches workshops on the farm on a variety of farming and homesteading topics. Carl's background is in forestry and draft animals, and he consults and teaches workshops on sustainable farm and forest management. Integrated into this mix of horticulture, livestock, forestry, teaching, and consulting is some very well-planned grazing management.

The grazing system Lisa and Carl have set up for their small raw milk dairy is a creative design that allows them to graze a very small group (usually two or three cows) on several different pieces of land during the grazing season. Despite the very small herd size, they have designed a grazing system that allows them to use a medium to high stock density. This is done using flexible paddock sizes and strip grazing, which allows them to control stock density by changing how often the front and back fences are moved. They can provide almost all the feed to their small dairy herd from pasture during the grazing season, while also using the cows to improve the pastures. They sometimes rotate other livestock, including pigs, horses, and poultry, through the pastures as well.

Lisa sets up the grazing strips so that each one lasts about two days. The front fence is then moved forward several times a day to encourage to cows to keep eating the high-quality plants while trampling the weeds and plant residual.

Each pasture area has a permanent perimeter fence. The interior subdivisions are created using portable

YEAR 2014			GRAZING PLAN		
	PADDOCKS		April	May	June
	Size	Number/Name			
Gilead Rd	1.8	EW3 Gilead			
Lg Garden	0.5	EW2a			
Horse Corral	0.25	EW2b			
Triangls	0.5	EW2c			
Garlic Fld	0.3	EW4 (garlic fld)			
Tom/Lisa	0.3	TL Corrall			
Tom/Lisa	0.3	TL Pond			
Tom/Lisa	0.3	TL Gravel			
Tom/Lisa	1.5	TL Flats			
Tom/Lisa	0.5	TL Slope			
Hntg Cbn	2	Phillips3			
Near T/L	1.5	Phillips2a			
Near Ph3	1.5	Phillips2b			
Renovated	1.5	EW1b			
Pasture	1.5	EW1a			
	1.5	EW1c			
	1	EW1d			
	1.7	Raimondi-a			
	1.7	Raimondi-b			
	1.7	Raimondi-c			
Log Home	1	EW5			

KEY
Horses — Foliar Sprays — Super Trace — Winter Feeding — Steers
Cows — BD F&G — Cows
Steers — Solubor
Clipping — Manure/Lime

Figure 10.2. Lisa and Carl do Holistic Planned Grazing and keep both their grazing records and grazing plan on their grazing chart. They also record fertility inputs and winter feeding locations on their chart. You can see a photo of a paper version of a grazing chart in figure 7.7.

Figure 10.3. Lisa McCrory moves a front fence by winding up the polywire on a low-cost electric wire reel. The cows enthusiastically walk into the new strip to graze a mixture of legumes and grasses including bromegrass, which thrives because of the taller pregrazing heights and longer regrowth periods that Earthwise Farm employs.

Figure 10.4. These portable posts and polywire reels are ready for use in setting up the next strip of pasture for grazing. The black plastic pipe is a semipermanent supply line for drinking water.

posts and polywire. This gives Lisa flexibility to change the shapes, sizes, and locations of each paddock. The portable interior paddock fencing can also be moved out of the way to make it easier to clip weeds or nontrampled plant residual when needed.

Because the pasture areas are small fields separated by woods and roads, Lisa and Carl use a portable solar-powered fence energizer. This is connected to a permanent grounding system built at each pasture location. A small milking area has been set up at each field so the cows don't have to walk a long distance for their daily milking.

All the pastures have drinking water available. Water is collected in a cistern at the top of the hill close to where the farmhouse is and then piped downhill using gravity so that portable tubs can be filled in each pasture. Cows also have regular access to a salt mineral mix and kelp in each pasture.

Forage quality has improved significantly in the last few years due to good grazing management, and the cows are from farms that have good zero-grain genetics. However, Lisa prefers to feed a couple of pounds of grain to each cow during milking, which allows the cows to make more milk and have better reproductive performance. The supplemental grain also helps balance out any variations in forage quality. Lisa and Carl monitor cow body condition, rumen fill, and milk production — as well as observing cow

Figure 10.5. This homemade fly trap, along with portable water tubs and a mineral feeder, is moved with the cows from pasture to pasture.

manure — to make sure the cows are healthy and getting enough of the correct feeds from the pasture.

Lisa and Carl have worked hard to build soil health and biological activity on the farm with the use of planned grazing, organic nonsynthetic fertilizers, compost, and biodynamic preparations. The large population of dung beetles in their pastures is one example of their success.

CHAPTER 11

100 Percent Grassfed

What are the reasons to feed a 100 percent forage diet to ruminants on a grass farm? One important benefit to farmers is that there is no expense or labor of feeding grain, so the potential profit can be higher. An increasing number of farmers and consumers are interested in the nutritional and health benefits of grassfed meat and milk. There is also the personal satisfaction that many grass farmers enjoy, of having a farm that is self-sufficient in livestock feed and has a smaller carbon footprint. If the forage is of sufficient quantity and quality to supply all the livestock needs, then the only inputs for the herd or flock will be minerals and perhaps some health care materials.

Providing 100 percent of the feed from forage, however, requires the right animal genetics as well as

Figure 11.1. Harrier Fields Farm in New York State has worked hard to find beef cows and bulls that grow rapidly and produce high-quality grassfed beef. This bull is an excellent example of a beef animal well adapted to a high-forage diet.

high-quality pasture and stored feeds. The farm manager must be a highly skilled manager. When it works, 100 percent grassfed is great, but a poorly managed system can result in poor animal performance and even animal welfare concerns.

Matching the forage quantity and quality to the livestock needs requires careful management to ensure there will be enough pasture and harvested stored forages available to meet all the nutritional needs of the animals for the entire year. For a beef herd or sheep flock being raised to produce meat, doing this without supplemental grain can be relatively easy as long as the farm has enough land producing quality forages, and the herd has reasonably good grass-based livestock genetics.

Producing milk, beef, lamb, or goat on 100 percent forage requires high-quality pasture, the right livestock genetics, and good skills in both grass farming and animal husbandry. The higher the performance level and productivity of the livestock, the higher the quality of the pasture and skills of the manager need to be. In addition, unless you are in an area where the climate makes year-round grazing possible, you will need stored forage as the primary feed source for part of the year. The nutritional quality and digestibility of that stored forage will also need to be high enough to meet nutritional needs.

Good management can produce forage of high quality that can meet all the needs of ewes and lambs, cows and calves, and even feeders being finished on pasture. Even with sheep, though, using the right genetics is essential. Trying to use genetics from a flock that has been bred for show sheep characteristics, or from several generations of grain-fed sheep, can result in disappointing grass-finished lambs. And it is a much greater challenge to produce the very high-quality forage required to meet all the needs of dairy cows, because modern dairy breeds have such high nutritional needs.

Most commercial grain mixes, pellets, or mash blends contain a mineral mix, so when replacing grain or a total mixed ration with pasture, it is important that the animals are still getting the minerals they need. In addition, it takes more than a pound of forage dry matter to replace each pound of grain dry matter in the ration. So animals switching from a high-grain

Figure 11.2. This lamb is from a closed flock of 100 percent grassfed sheep, where one of the primary selection criteria has been production of lambs that can finish on forage.

ration to a high-forage ration will need a lot of highly digestible forage. This may require adding additional acres of pasture and cropland for forage production. Once grain is no longer being purchased and imported onto the farm, fewer off-farm nutrients are passing through the animals to provide fertility to the pastures and crops. To compensate for this, it may be necessary to purchase additional fertilizer to maintain crop yield and quality. It will also be essential to monitor livestock health and performance (as described in chapter 9) to make sure they are adapting well to a zero-grain system. Monitoring for a beef or sheep producer should include tracking growth rates, forage intake levels, and body condition. For a dairy producer, monitoring should include watching milk production, forage intake, and paying careful attention to body condition and reproductive performance.

The level of difficulty with the transition to zero grain depends on the quantity of grain the animals have been accustomed to, their genetics, and the quality of the forage available. For example, a beef farmer who already has a zero-grain ration for cows and calves, and currently provides grain only to finishing steers, will have a much easier transition to 100 percent grassfed than a farm where all animal groups are fed grain.

Unique Challenges for Dairy Farms

For dairy farmers, making milk without any grain supplementation is possible, but it is also very challenging. In order to succeed with a zero-grain (100 percent grassfed) dairy, the quality of the forage, selection of dairy cows, monitoring of herd health, and production and skills of the manager must be exceptional. In addition, the quantity of forage needed is higher than expected, because the cows have to eat more of it to meet their nutritional needs.

Since most breeds of dairy cows have been selected to produce a high volume of milk, and most have been bred to do well on a ration which includes both forages and grains, many dairy herds may do poorly on a zero-grain ration. This is particularly true of the modern dairy breeds and animals selected and raised on farms that feed high levels of grain to both calves and cows. Because of these challenges, the remainder of this chapter will focus on dairy animals. However, much of this information can be applied to all types of ruminants on a high-forage zero-grain ration.

Recent History

Before modern dairy genetics were introduced, many cows produced milk without being fed any grain. However, modern dairy breeds are very different from older ones. The older breeds produced a lot less milk than cows make today, but they were able to thrive on a forage-only diet.

Because zero-grain dairy farming is very challenging, and has often been less profitable than adopting modern dairy breeds and feeding systems, until recently relatively few farms have been doing it successfully. During the 1990s, there was interest in zero-grain dairy production

in a few regions. Some farms found it successful, but others found that their cows did not do well, and that milk production was too low to cover farm overhead costs.

Some successful zero-grain dairy farms still in operation carefully transitioned to zero grain a decade or more ago, and continue to find that it works well for them. Other farms made the transition to zero grain more rapidly, due to financial pressures caused by high grain costs and, in some situations, unpaid grain bills. As this book is being written, a number of new farms are in the process of transitioning to zero grain so they can receive the organic "grass milk" premiums being paid by some organic milk companies. These farms are working hard to learn how to maximize dry matter intake from forages, maintain enough milk production to pay farm expenses, and monitor heard health and performance. As more farms learn how to raise dairy animals without grain, they will be able to share this information. Over time, this will help other farmers make informed decisions on how to make the transition successfully, or determine if zero grain is a good match for their farm at all.

Economic Challenges

The financial impact of cutting the grain out of the ration varies from farm to farm, based on pay price and the farmer's ability to maintain enough milk production and herd health without it. Most farms see a drop in milk production on a zero-grain ration. It is important to plan ahead of time for this drop in milk production and decreased farm income.

The savings by eliminating the grain bill may be offset by the need to make or buy more forage. Once cows must get all their nutrition from stored forages and pasture, they will eat a significantly larger amount of forage than they did while on a diet that included grain. Many farmers have found that in the transition to zero grain they needed to either buy more forages, add more land for both grazing and haying, or downsize the herd to better match the size of their land base.

Feeding high-energy grains such as corn or barley is the easiest way to create the ideal energy-to-protein ratio in a cow's rumen. High-quality pasture, which can be very high in protein, often exceeds the animal's protein requirements. Supplementing with a small amount of an energy source can make a big difference to rumen efficiency, body condition, milk production, and weight gain. For farms with modern dairy genetics, which have not been selected to do well without grain energy sources, the benefits of feeding even a small amount of high-energy grain can be significant.

In response to increasing consumer demand for grass-fed dairy products from cows fed only forages, some milk buyers are now paying a premium for milk from cows fed zero grain. In addition, many dairy farms that are direct marketing raw milk or farmstead cheeses are able to set their own prices, and are selling to customers who are increasingly interested in grassfed dairy products.

Each farm will need to evaluate the financial pros and cons of eliminating grain. For some farms, the grassfed milk premium may make it financially worth-while. However, for other farms, the premium may not be enough to cover the loss of milk income due to lower milk production, potential poor reproductive performance of the herd, and increased need for more forage on the farm. As consumers become more willing to pay extra for zero-grain dairy products, the economics of producing them will hopefully improve for dairy farmers.

Farms that transitioned to zero-grain feeding systems long before any 100 percent grassfed premium was available obviously did it for reasons other than higher premiums. Zero-grain farmers will speak of the psychological benefit of not facing a monthly grain bill. Many say they feel good about being totally self-sufficient in feed. Others are motivated by the potential nutritional and health benefits from eating grassfed products. Some of the farms that do grow grain find that they prefer to feed it to nonruminants (pigs and chickens) or sell it for human consumption. So while economics are important, some continue to use little or no grain even without the premiums.

Monitoring and Management

A zero-grain dairy herd will eat more forage, require more acres of pasture, and need a good supply of minerals. In order to prevent herd health and production problems, monitoring performance, production, and health of the herd is also essential.

LAND FOR FORAGE PRODUCTION

Many successful zero-grain dairy farms reduced the herd size as part of their transition, or they added acreage of both pasture and harvested forages. Farms that have successfully transitioned to zero grain also report that they added more *good*-quality cropland, not just more hilly or marginal pastureland.

The increased need for acreage is due to:

- The need for more acres of pasture per cow to increase pasture dry matter intake per cow.
- The need to cut hay earlier (resulting in lower first-cut yields), which is important to supply forage that is less mature, more digestible, and higher quality.
- The need to feed more stored forage per cow to replace dry matter the herd is no longer receiving in the form of grain.

Many farms report that they feed some stored forage during the grazing season in addition to giving the herd more acres of pasture each day. This encourages the cows to come into the barn to be milked, and adds some extra dry matter to their diet. They also allow the cows to waste some pasture in order to get as much pasture dry matter into the cows as possible. The extra pasture left behind was either trampled (on farms using high stock density) or clipped after grazing (on farms with lower stock density).

FORAGE QUALITY AND QUANTITY

Though the non-grazing season isn't the focus of the book, zero-grain farms do need to plan ahead for the months when stored forages are fed. Feeding highly digestible forages so that the cows can eat as much forage dry matter as possible is a priority during these non-grazing-season months. This generally requires feeding second or third cut, and some farms also take an earlier first cut — which although lower yielding has higher digestibility.

Running out of high-quality forage partway through the winter can force a switch to feeding forage with lower digestibility. Without the availability of grain in those

Figure 11.3. While waiting to walk through the milking parlor, the cows at Spring Creek Farms are fed haylage. Notice the herd's good body condition, full udders, and excellent rumen fill even though they are not being fed any grain. See the farm profile at the end of this chapter to learn more about this farm.

situations, cows may suddenly make significantly less milk or lose body condition.

Soil Fertility Inputs

Since nutrients are no longer being imported onto the farm in the grain truck (and then passed through to the fields in manure), it may be necessary to use more off-farm fertility inputs to improve forage quality and yield. Some of the companies that buy zero-grain milk pay extra to encourage farmers to spend more on soil fertility inputs and improved forage quality.

In addition to fewer nutrients entering the farm in the absence of grain purchases, the farm actually needs to produce more tonnage of forage. If this results in more acres of land per cow, there is now less manure to spread per acre. So this can compound the fertility shortfall on some zero-grain farms.

In order to prevent this, farms managing for high-forage yields and quality are using a variety of strategies

and fertility inputs. These include some crop rotation with annuals, use of chicken manure or cow manure from other farms nearby, and purchased fertility such as wood ash or lime. Any of these inputs needs to be based on soil test results and assessing what is growing and current yields. Refer to chapter 4 for more information on pasture soil health.

Reproductive Performance

Poor reproductive performance can be a common problem in the first few years of transitioning dairy cows to zero grain. This can result in gradual reduction in milk per cow as the whole herd shifts to being mostly in late lactation, with a larger-than-ideal number of open cows. Several of the successful long-term zero-grain dairy farms have selected their heifer replacements based on reproductive performance. Over the years they have kept only heifer calves out of cows that bred back quickly after calving. By doing this they have selected for genetics of animals that can maintain body condition and breed back quickly without supplemental grain.

Zero-grain dairy farms that transition their herds off grain rapidly, without the right forage quality or cattle genetics, may end up with a significant number of cows taking too long to breed back (or not breeding). This can create a serious cash-flow issue while you wait for the cows to get bred back, calve, and allow fresh cows to enter the herd, making more milk. Monitoring average days in milk of the whole herd, the average calving interval, and the number of days between calving and successful breeding will help catch this issue early.

Minerals and Energy Sources

Since the livestock on a 100 percent grassfed farm aren't receiving grain, which usually contains additional minerals, you will need a new method to supply minerals and salt. Most zero-grain farms use a loose mineral salt mix or free choice minerals instead of solid lick blocks to make sure livestock are able to get enough: It's easier for livestock to eat the loose mix quickly than to lick a solid block.

If the balance of energy to protein is a concern, it may be possible to add some non-grain energy sources to the ration, depending on the rules set by the milk buyer or grassfed standards. Some farmers are feeding molasses in both liquid and dry form, some high-energy root crops, and even sugar. However, some of the buyers of 100 percent grassfed milk set standards regarding which energy sources are allowed. Some farms will switch to milking just once a day as part of a strategy to make less milk and decrease the distances the herd has to walk to the milking area each day. As cows walk less, and make less milk, it can be easier to meet their nutritional needs with just forages.

Adjusting to Lower Milk Production

Milk production per cow per year on zero-grain dairy farms can run from as little as 4,800 pounds per cow to as much as 11,000 pounds. However, it appears that the majority of zero-grain farms produce from 6,500 to 8,000 pounds per cow.

Successful farms producing higher amounts of milk per cow generally have been using zero-grain rations for several years. During that time they have been making genetic selection decisions for cows that did well in the system, as well as improving forage quality. Many zero-grain dairies milk just once a day during some times of the year as part of their strategy to match forage intake to milk production.

Not all farms have found that the lower milk production levels worked for them financially. Even with the elimination of the grain bill, there needs to be enough income left to cover farm overhead costs and increased forage needs. Each farm needs to investigate the challenges and decide whether a zero-grain system could work in their unique situation. This needs to include agronomic, animal, and financial considerations. Farms that lack enough land for pasture and hay, or that have high overhead costs, may find that a zero-grain system will not work for them.

Managing for Success with Zero Grain

The number one principle for success with zero-grain dairy rations is managing to maximize forage dry matter intake (this is discussed in detail in chapter 9). However, the approach that each farm takes to accomplish this goal will vary greatly. Good genetic selection and feeding a lot of high-quality forage are also important, but there is no simple recipe for why some farms find that a zero-grain system works and others don't. Each farm will need to find its own best system, and not all farms will find that it is a successful management system for their unique situation.

- Maximize dry matter intake of forages. This requires highly digestible forages.
- Cows will consume a lot more forage to replace the grain they are no longer eating.
- Make sure you have enough land for increased forage consumption from crops and harvested forage.
- Monitor body condition to make sure the herd isn't losing too much weight.
- Monitor reproduction to make sure cows are breeding back on time and staying bred.
- Select for the right herd genetics.
- Make sure that the level of milk production brings in enough income to keep the farm financially sustainable.

The Art of Good Grazing

Zero-Grain Dairy

Forrest Stricker and his son Greg have a dairy farm in Pennsylvania where they feed 100 percent forage — zero grain. During the grazing season, the goal on Spring Creek Farms is to keep the cows healthy and productive while getting as much dry matter from pasture as possible.

Over the years the Strickers have used many methods to improve the diversity and overall productivity and quality of the pasture. In the past few years, they have been reseeding some areas with improved cool-season perennial forage species. They reseed by frost seeding, rotating through an annual crop, or using a GenTill to add seed into perennial sods. But the biggest management change they have made recently has been to let the pasture grow a bit taller before the cows graze, and to leave a little more residual in each paddock. The excess forage the cows don't graze is either trampled by the herd as they graze or clipped with a mower after the paddock has been grazed. With the recent management changes, the Strickers are seeing steady improvement in the pastures.

The cows are in good condition and make over 9,000 pounds of milk per cow per year on a diet of pasture, hay, and a tiny amount of energy supplement. The herd gets a fresh new pasture area at least twice a day, but may sometimes get moved to new pasture four times a day. They sometimes feed balage or dry hay in the barnyard when cows come in to be milked each day. However, during the 6 to 10 weeks each year when pasture quality and quantity are at their highest, there is no need to feed any additional forage. Cows also get an energy supplement at some times of the year. This is fed only when the pasture protein levels are high and the cows need some extra energy. The cows often have to walk long distances to and from pastures several times a day. Using high-quality pasture and not pushing the cows to graze too hard makes the walks easier for them.

The Strickers constantly monitor the herd by watching to make sure rumens are full, the manure isn't too loose, body condition is good, and milk production

Figure 11.4. The Strickers' milking herd has finished grazing this paddock and is waiting to walk back to the parlor for afternoon milking.

is on target. Forrest says that he and Greg constantly balance the pasture and forage quality with the level of milk production, body condition, rumen fill, texture of the manure, and milk urea nitrogen levels. This allows them to catch problems early, and adjust grazing and supplemental feeding regularly.

The grazing season generally starts in early April and goes into late November or December depending on the weather. In spring, when the pastures are growing fast, the herd is moved on a fairly rapid 21-day rotation through somewhat larger paddocks. This allows the cows to eat the more digestible legumes and vegetative grass (the leaves). The cows eat some of the emerging seed heads and stems, too, but are never left in a paddock so long that they are forced to eat those less digestible plant parts. Later in the spring and summer as the pasture growth rates slow, the rotation is slowed to 30 or 40 days through a combination of reducing paddock sizes and adding more acreage to the grazing rotation.

The farmers usually clip each paddock in the dairy pastures once a year unless the residual is heavily trampled after grazing. Clipping or trampling the residual encourages the cool-season grasses to grow vegetatively for the rest of the season, and removes the coarse standing stems that can slow down the herd's intake of highly

Figure 11.5. By using a taller pregrazing height, leaving more trampled plant residual behind, and letting some perennial grasses produce seeds, Forrest Stricker has created a pasture understory environment where new grasses can germinate and grow even without reseeding.

digestible pasture. This clipping also cuts off the weeds before they can make seeds.

The farm is rolling hills, with cow lanes heading out in many directions from the barn. The lanes and perimeter fences create large, unevenly sized (and shaped) pastures, which are each subdivided into different-sized paddocks with polywire depending on the number of cows in the grazing group and the amount and quality of pasture. Black plastic pipe runs along many of the fence lines and lanes to provide piped water to most areas of the farm.

In addition to the milking group of animals, there is a group of nurse cows with calves and younger heifers, which graze another area of the farm. There is also a group of older heifers and dry cows with the bulls. Each of these groups is given fresh pasture once or twice a day. They often graze areas of the farm that are too far from the parlor to be convenient for the milking group to walk to.

Figure 11.6. The nurse cow heifer group is grazing a new seeding of alfalfa, perennial grasses, and puna chicory. This area is being grazed with a high stock density using strip grazing.

Figure 11.7. Greg Stricker checks the mobile hen house, which is moved into pastures where the cows have recently grazed.

The plant species on the farm include red clover, white clover, alfalfa, orchard grass, tall fescue, meadow fescue, perennial ryegrass, Kentucky bluegrass, quack grass, and narrow-leafed plantain; several fields include some puna chicory. Now that Forrest and Greg are managing the pastures to have a taller pregrazing height, they may try to plant some smooth bromegrass in the hope that it will boost pasture productivity during drought conditions. Brome is one of the elongating types of cool-season grasses that does best when given longer regrowth periods and not grazed too short or too often.

In addition to the cows, this farm also grazes two flocks of chickens behind the cows in mobile chicken coops. Each coop is guarded by a dog. Eggs are sold in the on-farm shop, where they also sell beef, raw milk, cheese, and butter. Customers can watch the cows come in from pasture and be milked in the parlor right next to the shop.

CHAPTER 12

Health Concerns

Health concerns specific to livestock on pasture can include bloat, metabolic disorders, poisoning, predators, and parasites. Farmers can prevent all these potential problems, or at least keep their incidence low, through good management, monitoring of livestock and pastures, and thoughtful planning.

Bloat

Bloat occurs when rapid fermentation in the rumen creates high levels of gas. If the excess gas is not released fast enough, the rumen will "bloat" and expand and may kill the animal due to pressure on other internal organs.

Most legumes other than bird's-foot trefoil have the ability to cause bloat in some situations. Bloat is most likely to occur when livestock graze frosted legumes, although it may also rarely occur when livestock are grazing lush small grains early in the spring. Grazing alfalfa when it is wet from dew or rain in the morning creates a higher risk of bloat compared with waiting until the plants have dried.

Not all legumes pose the same level of bloat risk. Growing and grazing more bird's-foot trefoil can reduce the risk, particularly in pastures intended for fall stockpiling. Additionally, planting bloating legumes in a mixture with grasses and forbs will allow livestock to graze a broader range of plants, cutting risk.

To reduce the chances of bloat, don't turn livestock that are hungry into fields that contain a lot of wet or recently frosted bloating legumes. Some farms report that providing salt and minerals may decrease bloat risk, but research to date doesn't back up this idea. Other strategies proven to reduce risk of bloat include administration of preventive treatments that include active ingredients such as poloxalene, and feeding some stored forage just before turning animals out onto risky pasture.

Poloxalene-containing treatments are currently allowed even on USDA-certified organic farms.

Treatment of acute bloat is best done by a veterinarian. However, because it can come on quickly and be fatal, it is a good idea to develop a treatment and prevention plan through conversation with the farm veterinarian ahead of time. Keeping a supply of a treatment product on hand may be a good idea if there is a risk of bloat.

Grass Tetany

Grass tetany is caused by a low blood level of magnesium (Mg) and can lead to collapse and death if not treated quickly. Symptoms — including extreme nervousness, muscle spasms, and awkward gait — may come on very quickly and progress rapidly to death in some cases. This serious problem is also called grass staggers, magnesium staggers, or winter staggers in different parts of the country.

Although this can occur any time of the year, it is most common when livestock are grazing lush spring pasture. It is also more common in grassy pastures with a low legume content. If soils contain excess levels of potassium, low levels of phosphorus, and are fertilized with nitrogen in the spring, they may produce forages more likely to cause grass tetany.

Dairy and beef cows that are about to calve or have recently calved are most susceptible. Providing a mineral mix that contains magnesium can help reduce the risk of grass tetany. When grazing pastures that are mostly grass, wait until plants are 8 to 10 inches tall. Increasing the amount of legume in the pastures will also lower risk of grass tetany.[1]

All animals showing signs of grass tetany should be removed from the pasture and fed a good-quality legume/grass hay along with minerals containing magnesium. Treatment of a serious case of grass tetany is generally done by a veterinarian, and involves either oral or IV magnesium supplementation. Oral magnesium solution can be made with Epsom salts and water.

Selenium Deficiency

Health problems caused by selenium deficiency are not specific to grazing, but these illnesses can be a serious concern on farms where 100 percent of the feed comes from homegrown pasture and forages grown on selenium-deficient soils. Symptoms of selenium deficiency include weak or dead calves, reduced immune function in older animals, cows retaining placentas after calving, and lambs or kids that appear "stiff" or unable to walk (white muscle disease).

Treatment of severe cases is challenging, so prevention or early treatment is important. For farms with low soil selenium levels, it can be added to the salt and mineral mix. Selenium can also be provided by injections. However, selenium can be toxic if livestock get too much. Consulting with your veterinarian or nutritionist is the best way to assure livestock are getting enough — but not too much — of this essential mineral!

Poisoning and Toxicity

Poisoning or toxicity problems in livestock can occur on pasture due to animals eating foreign materials in the pasture, poisonous plants, or forage species that have toxicity issues in some situations. These sorts of problems are more likely to occur when new animals are brought to the farm or pasture, or if animals are hungry and searching for food.

Hardware Disease

Hardware disease is the name given to the symptoms shown by animals who become ill after eating foreign objects such as wire, nails, or other metal pieces. These objects usually go into the reticulum, where they may puncture and cause infection or damage to nearby organs. Symptoms of hardware disease may include stiffness, lack of cud chewing, and loss of appetite.

Prevention obviously calls for making sure there is no old wire or trash in the pastures. However, this can be challenging if you're taking over a farm that previously used barbed-wire fences or had trash or burn piles in pasture areas.

Treatment in some cases can involve the use of a magnet placed in the stomach. However, more serious cases may require surgery and other veterinary treatment.

Toxic Plants

Plant-specific toxicity actually occurs less frequently than you'd expect based on the large numbers of livestock grazing in areas with toxic plants. As discussed in chapter 10, under good, consistent management, animals learn not to eat common toxic plants.

In *Foraging Behavior*, Dr. Fred Provenza wrote:

> When herbivores forage, however, over-ingestion of toxins is seldom a problem. Rapid postingestive feedback from toxins enables animals to limit the rate and amount of most toxic foods ingested, apparently in accord with the rates of detoxification they can sustain. Thus, the concentration of toxins in foods sets limits on the amount of a particular food animals can ingest. As toxin concentrations in a plant decline, intake of the plant increases. That is why, given a choice, herbivores are able to select more of foods that are high in macronutrients and low in toxins.[2]

Still, some plants are acutely toxic, and — particularly for animals that have just been moved to a new farm with plant species they're not familiar with — poisoning and death may occur. Some examples of plants that can be acutely toxic include false hellebore, deadly nightshade, and the wilted leaves of cherry trees.

Other plants that contain toxins may be consumed in small amounts without making the animals ill, or at least hopefully not too ill! Some toxins such as nitrates or prussic acid are only present in plants at certain times of the year or under certain growing conditions. Forage species that have potential toxicity issues include tall fescue, sorghum sudan grass, and brassicas.

Prussic acid poisoning. This type of poisoning happens when levels of cyanide-containing compounds accumulate in certain plants. This is primarily a risk under certain growing conditions with plants such as sorghum sudan grass and Johnsongrass (see the sections on these plants in chapter 6).

Conditions that can increase the risk of prussic acid poisoning include allowing livestock to selectively graze with a low stock density. This allows them to select and eat more young leaves, which are higher in prussic acid. Grazing with a higher stock density forces the animals to eat more stems and a variety of mature and less mature leaves so the overall level of prussic acid in their diet is lower. Prussic acid levels are also higher in newly frosted leaves, in regrowth after a frost, and during extended drought.

If you're planting forage species that carry a risk of this problem, select varieties with low prussic acid levels. Another strategy to reduce risk includes carefully following fertilizer application recommendations and not overapplying fertilizers. In addition, graze plants such as sudan grass or sorghum sudan grass hybrids after they are at least 18 to 24 inches tall, and take extra precautions if plants are drought stressed or have been frosted.[3]

Nitrate poisoning. This type of poisoning is most likely to occur when nitrogen fertilizer has been applied followed by drought conditions. It may also be a risk if excess manure or fertilizer has been applied to a pasture. Forages which present a risk of nitrate toxicity can be tested by a forage testing lab to help determine whether they are safe to feed livestock.

Fescue toxicity. The cause of this problem is a fungal endophyte that infects tall fescue. Symptoms of fescue toxicosis can include reduced weight gain, lower fertility, reduced feed intake, elevated body temperature, and a condition called fescue foot. Fescue foot is caused by reduced blood circulation to the extremities of the animal's body, so that tips of the tail, ears, or feet can be damaged or necrotic.

Fescue toxicity is best avoided by not planting endophyte-infected forage varieties. If pastures are already full of infected fescues, diluting them by interseeding with clover or other legumes can reduce risks. When grazing infected fescue pastures, careful monitoring of the livestock is important, particularly during times of the grazing season when endophyte levels are the highest.

Brassica-related problems. Under certain situations, brassicas (such as turnips and kale) can cause some health problems if not grazed correctly and carefully. Problems can include bloat, nitrate poisoning, atypical pneumonia, hypothyroidism, and hemolytic anemia. To prevent these issues, grazing animals must be introduced slowly to brassicas, generally over four to five days. In addition, brassicas should be used to provide only a portion of the animals' diet. Make sure the herd or flock is not hungry when turned into the brassica field. Strip grazing is a common way to carefully graze brassicas and limit daily intake by livestock. This can be done by providing supplemental forage in the same pasture but limiting their access to the brassicas, or by grazing them in the brassica strip for a limited time period each day and then moving them to graze other forages.

Internal Parasites

Internal parasites live in all grazing livestock. In healthy adult cattle, they generally won't increase to a level where they can cause problems. However, in sheep, goats, and young dairy or beef cattle they may become a serious problem. One of the most common sources of parasite infection is pastures, particularly when they are poorly managed or overstocked.

Farms with grazing systems that encourage livestock to graze the pastures down very short are more likely to have parasite infection problems. In addition, farms with more grazing animals per acre of pasture will be more likely to struggle with parasites. By learning more about the parasite lifecycle, and all the methods available to prevent and treat parasites, a grass based livestock farm should be able to develop an effective parasite prevention plan as part of the grazing system.

THE PARASITE LIFE CYCLE

Many of the internal parasites of ruminants have a somewhat similar life cycle, as shown in figure 12.1. The cycle of infection begins when eggs or larvae get deposited in the pasture in manure. These parasites continue to develop to the infective stage and are then consumed by grazing animals. This infects the new animals, and leads to the development of a new generation of parasites. Along the way, the parasites can cause health problems and illness or even death in the livestock.

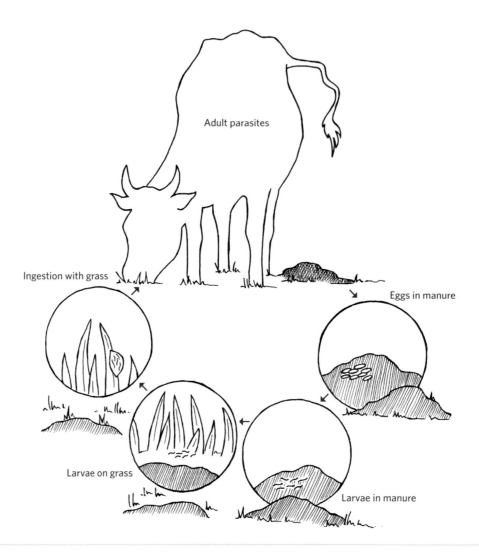

Figure 12.1. The cycle of infection begins when eggs or larvae move from livestock to pasture in the manure. Once in the pasture, the parasites continue to develop to the infective stage and are then consumed by animals as they graze. This leads to the development of a new generation of parasites in the host animal. Illustration by Anna Powell.

Not every parasite follows the same life cycle. Some have an indirect life cycle, which means that they spend a part of their life in a different or intermediate host species. The liver fluke, for example, spends some of its life in snails before infecting ruminants. Keeping animals healthy and productive during the grazing season requires a good understanding of parasite–host interactions, grazing management strategies, livestock nutrition, and animal selection.

Prevention and management of parasites so they don't become a health problem for grazing livestock will vary depending on the type of livestock, their level of susceptibility, how good their nutrition is, what the local parasite species is, and the local climate. Weather has an effect on how long parasites live and remain infective in pastures. High humidity allows parasites to remain infective in pastures for much longer. Hot and dry weather will reduce the number of infective parasites in pastures.

TESTING FOR PARASITES

There are several methods of testing or monitoring livestock for parasites. This is particularly important for farms raising those types of livestock that are most susceptible to illness from parasites. By testing, problems can be detected before animal health is negatively impacted. Testing provides useful information on which animals may need immediate treatment or veterinary care, and can also tell you if your prevention plan is actually working! Test results can also help with decisions on which animals to keep in the herd and which animals should be culled.

Two common testing methods are fecal egg counts and FAMACHA testing. Each can provide useful information on the health of the animals and how well the prevention and treatment system is working. Information on how to find labs, workshops, or training in both these methods can be found in appendix E.

Fecal egg counts can be done by your veterinarian, though some farmers choose to set up their own equipment and get trained to do some of this testing themselves so they can get frequent information on parasite levels in their animals. There are also labs that you can mail samples to. This test tells you how many worm eggs are being excreted in the manure — information that can be helpful in many ways. Fecal egg counts can help us understand what an animal's parasite load is, what seasonal changes are occurring in the parasite population, and how effective our prevention or deworming strategy is. This information can also help with decisions on which animals to keep, and which to cull.

FAMACHA is a method used on goats and sheep. This system was originally developed in South Africa but is becoming increasingly popular with small ruminant farmers in the United States and Canada. This method uses the eyelid color of an animal to provide an estimate on the level of anemia associated with barber pole worm (*Haemonchus contortus*) infections in sheep and goats. Training in this method is required, and there is information in appendix E on how to find the closest workshop or instructor.

Prevention and Treatment

Some parasites can be prevented through good sanitation, while others can be more challenging to control. It is beyond the scope of this book to discuss the many different parasite species, the illnesses they cause, and the prevention and treatment strategies required for each one. Work with a veterinarian to create a parasite prevention and treatment plan including good pasture management. The farm profiles at the end of chapter 3 and this chapter include some real examples of farms with grazing systems and parasite prevention plans that are working well.

Due to the fact that parasites are developing resistance to multiple chemicals, relying on synthetic dewormers as your only control method isn't a great idea! Even non-organic farms should be looking at alternatives to routine deworming with synthetics. These synthetic dewormers, when used carefully, can continue to be an important management and treatment tool for farmers. However, misuse can result in increasing levels of parasite resistance, which will decrease their efficacy further. Misuse of some of these dewormers can also kill dung beetles and other soil organisms that are essential parts of a healthy pasture ecosystem. If you're able avoid the use of synthetics or use them only strategically, your livestock, ecosystem, and income will all benefit.

Many types of natural, nonsynthetic dewormers are available. While some farms attribute their success with parasite management to these products, there is limited research on their real effectiveness. Many of the farms seeing fewer parasite problems may actually be seeing the positive results of good management from improved nutrition, better genetics and selection, and increased pasture quality. So it is not easy to confirm how much of their success is due to these treatment materials versus other management improvements. Still, good research has been done and is continuing on nonsynthetic products. Hopefully, this increasing knowledge base will help provide more options to farmers in the prevention and treatment of livestock parasites.

Genetics, Selection, and Nutrition

Selecting animals that shed fewer eggs in their manure, or that have some tolerance or natural immunity to parasites, is important, particularly for small ruminant farmers. Animals can be selected for resilience, which allows animals to maintain productivity despite infection. They can also be selected for resistance — the ability to prevent or limit infections. Continuing to select for animals that shed the fewest parasite eggs, and that do well compared to others in the flock, will lead to more productive livestock over time.

Beware of someone claiming to have a flock for sale that is completely immune to parasites! It often takes a few years for livestock on a new farm to show parasite problems, but as the stocking rate increases and more pastures and livestock become loaded with parasites, many of these supposed parasite-resistant flocks will also begin to show symptoms. So while selection and genetics are important, it is only one part of an overall parasite management strategy.

Maintaining good nutrition is also essential to make sure animals do as well as possible despite carrying a population of parasites. Animals lacking enough protein, energy, or essential minerals in their diet are more likely to become sick from parasites. Continuing to feed milk to calves, lambs, or kids past the normal weaning age can help them do better despite hosting some parasites in their digestive system.

Including forages that are high in condensed tannins can also help livestock with internal parasites. These plants include forage chicory, bird's-foot trefoil, and many of the browse or forb species (refer to chapter 6 for more details). Some farms graze the sheep or goats in areas of scrub or brush several times each season to provide them with additional high-tannin browse. The farm profile at the end of chapter 3 discusses a flock that gets browse in the grazing rotation.

Grazing Management

Since most of the parasites we are discussing are infecting the animals in the pasture, a carefully planned grazing system is an essential part of prevention. Most of the infective larvae are found in the lower few inches of the pasture, so leaving a post-grazing residual of at least 2 to 4 inches will minimize parasite infection. If the livestock don't graze the pasture too short, they are much less likely to ingest infective larvae. As discussed earlier, this practice is also better for the plants and provides better nutrition to the livestock. Thus, leaving a high post-grazing residual is a win–win plan for plants and animals.

Long rest periods allow plants time to regrow taller, so livestock are less likely to graze the plants down short. Planting or managing pastures to encourage taller pasture forage species will also make it easier to manage grazing to leave higher post-grazing residuals. In addition, if the weather is hot and dry during that longer regrowth period, some parasites may desiccate, which will reduce the number of living infective larvae in the pasture. However, simply rotating and resting the pastures in a humid climate will not be enough to significantly reduce the parasite population. Without a dry period, many parasites can survive for months or even more than a year in the pasture.

Forward grazing and a well-planned multispecies grazing system are two strategies that can reduce rates of infection. In forward grazing, livestock are regularly rotated to new "clean" pasture — areas that are not infected with parasites.

When planning in what order to graze different areas of the farm, consider which areas are most heavily infected and which areas are "clean" of parasites. For example, pastures will be more infective if animals that are shedding large numbers of eggs have grazed there within the last few months. Pastures that have been tilled and reseeded will be less infective, as will those that have been mowed or grazed short during a dry period. Keeping track of this requires grazing records of which groups grazed each pasture and when. Forward grazing works best when using a written grazing plan to ensure there is enough forage for the flock available in noninfective pastures throughout the whole grazing season.

Grazing with multiple species can be helpful in some situations. Goats can graze around cattle manure and cows can graze around goat manure — they are dead-end hosts for most of each other's parasites. However, simply putting cows and sheep together in a pasture isn't going to solve the problem. It is unfortunately much more complicated! For example, grazing every other time through the pasture with a herd of cows can reduce the level of infective parasites for sheep, but it still won't eliminate all the parasites in the pasture.

Diversity of plant species in the pasture can be another useful part of a parasite management system, because it provides more minerals and a wider selection of nutrients to the animals. Diversity also provides opportunity to include some high-tannin forages.

The Art of Good Grazing

Grazing Dairy Sheep

Vermont Shepherd Farm grazes several hundred dairy sheep and makes sheep milk cheeses. David Major and his family have been milking sheep, making cheese, and managing the flock on pasture for over 20 years now, and have learned a lot about how to improve pasture quality to keep both the land and livestock doing well.

Because small ruminants such as sheep are more susceptible to internal parasite problems than dairy and beef cattle, the grazing management strategy on this farm is designed to prevent parasite infection and problems. The flock is managed in three separate grazing groups to make managing the animals, parasites, and pastures easier.

There is one group of ewes that lamb in early spring. These ewes make up the main lactating group, and their weaned lambs are grazed in a separate area of the farm.

A second group is made up of ewes bred to lamb later in the summer. These ewes are added to the lactating group later in the grazing season.

The farm's parasite prevention system has been very successful, so deworming does not have to be done routinely and the livestock are healthy. However, some groups may still require the use of a small amount of dewormer to remain in optimal health and avoid reinfection. This very limited use of dewormers has many benefits, including protection of soil life such as dung beetles, reduced cost, and consumer confidence that the farm is not overusing synthetic chemicals.

The lambs are the group most susceptible to internal parasites, so they are rotated into pasture that was previously hayed or otherwise not grazed by sheep in the current year. This allows the lambs to graze pastures that are mostly "clean" of infective parasite larvae. Lambs are moved frequently so they are not forced to graze the pasture down too short. This avoids exposure to infective parasites, most of which are found in the lowest 2 to 3 inches of pasture growth.

The ewes, which have better-developed immune systems than the lambs, are rotated through the pastures every three to five weeks during the grazing season. The strategy of using a taller pregrazing height, planning

Figure 12.2. Good grazing management with sheep has steadily improved the pasture quality and productivity on Vermont Shepherd Farm.

Figure 12.3. At the farm, ewes are moved at least twice a day into fresh pasture. They eat the most digestible plant leaves, leaving behind trampled residual and manure; this protects and improves soil health.

Figure 12.4. Net fence is connected to the energizer power source using a single strand of highly conductive polywire. This allows areas of the farm where there is no permanent fence to be strip grazed using electric net fence. This fence can quickly be moved out of the way so the area can be hayed when it isn't being grazed.

for a short period of occupation, and not forcing the ewes to graze too short reduces the risk that they will pick up parasites.

Grazing is done using portable fencing and strip grazing. By using very few fixed or permanent fences, the farmers allow themselves a lot of flexibility to either cut hay or graze most areas of the farm. Strategically grazing or haying is one of the ways they create "clean" pastures with fewer infective parasites.

Fences are moved at least twice a day to give lambs and ewes fresh grass. Back fences behind each group are moved forward less frequently, but at least every three days to prevent regrazing of plants after they have started to regrow. Drinking water is provided using portable tubs in each paddock, and there is piped water to all parts of the farm.

During the grazing season, ewes are fed a small amount of grain in the parlor. The grain is a high-energy mix to balance the high-protein pastures. The sheep and lambs are not fed any supplemental hay or other fiber during the grazing season. Instead, over the years the farmers have gradually increased the pregrazing height to allow a higher level of fiber in the pasture.

Plant species are cool-season perennial grasses and legumes, and the farmers have not reseeded recently. By using good grazing practices, the farmers have naturally developed high-density, highly diverse pastures that provide a large quantity of dairy-quality forage to the flock.

PART FOUR

Designing and Managing a Grazing System

Now that we have looked at grazing from the perspective of the plants and that of the livestock, we can see how their requirements complement each other synergistically. Plants respond best to short periods of grazing followed by long regrowth periods. Livestock do best in pastures that are managed so each paddock is grazed quickly and that are full of high-quality forage that has been given sufficient time to regrow. Good grazing management is a win–win system for plants and livestock. There are significant environmental benefits, too, including improved ecosystem resiliency, more sustainable farm income, human health benefits from grassfed products, and decreased use of potentially harmful agricultural chemicals.

With this solid understanding of the needs of *both* plants and livestock as the foundation, farmers can successfully design a new grazing system or improve an existing one. In this part of the book, the focus is on the practical aspects of designing a system, including how to assess land base requirements, calculate paddock sizes, construct effective fences, provide drinking water, and allow for efficient low-stress animal movement through the pastures. We will also discuss some of the details of management such as stock densities, managing multiple animal species, record keeping, and how to monitor pasture productivity and health.

CHAPTER 13

Pasture Math: Calculating Acreage

One fundamental question that a grass farmer needs to answer when designing a grazing system is: How many acres of pasture do I need overall? Having the right amount of good-quality land is necessary to design a sustainable, successful system that will provide enough feed for the herd while also doing a good job of taking care of the pasture plants' needs. In order to answer that question, though, the farmer first needs to know:

- How much pasture dry matter does the grazing group need?
- How much pasture dry matter is available on the farm for that grazing group to harvest?

The basic grazing principles of short periods of grazing followed by adequate recovery periods will come into play as you gather the necessary data and make calculations of how much acreage is required to support the dry matter needs of the herd or flock.

How to Measure Pasture

Chapter 8 explains how to figure out the amount of required dry matter intake or dry matter demand (DMD) for a particular herd or flock. So let's turn our attention here to the important task of measuring the amount of dry matter growing or available per acre.

Making an accurate assessment of how much forage a pasture can supply is critical, because underestimating the amount of forage available can result in a lot of wasted feed. On the other hand, overestimating the amount can result in a hungry flock or herd of animals when they run

TOTAL VERSUS AVAILABLE DRY MATTER

With any method of measuring pasture dry matter, it's important to understand the difference between *total* dry matter and *available* dry matter. Total dry matter is the entire amount of plant material in a pasture. Available dry matter is the amount of *grazable* material in a pasture. Since we aren't forcing the animals to eat every bit of forage in the pasture, it is important *not* to use total dry matter per acre amounts when making feed intake calculations. Some of the pasture dry matter measurement tools and methods give you the total pounds of DM per acre, while others give you just the amount that the animals will consume.

The difference between total dry matter and available dry matter varies from 30 to 50 percent, depending on the severity of grazing. Dairy cows, for example, do best when they are moved frequently

and allowed to harvest no more than half of a pasture's total dry matter. So if a herd of cows is moved into a new 1-acre paddock that has 2,800 pounds of total dry matter per acre, the goal might be for them to take in only about 1,200 pounds of that dry matter. In this example, they would be leaving 1,600 pounds of standing and trampled forage behind. In this situation, when making feed intake and paddock sizing calculations, you'd assign this pasture a value of 1,200 pounds of available dry matter per acre.

This approach reflects good grazing management principles. It is ideal for the plants, which don't like to be grazed too short. It is also ideal for the cows, since they prefer to eat the upper parts of the pasture plants rather than the more fibrous parts closer to the soil surface.

out of forage. In this situation, animal performance will suffer, and they are likely to graze the plants too short and damage them. They are also more likely to break out of the fence when they are hungry!

There are many methods of estimating or directly measuring forage dry matter per acre. These include simple measurements of height, or more complex measurements that also factor in plant density. The tools used include low-tech measuring sticks, grazing sticks calibrated to density and height, several types of plate meters, and an electronic pasture probe. Though it's tedious, dry matter measurements can even be done by hand clipping and weighing.

These tools vary in cost, accuracy, and difficulty, and any of the methods take some time and practice to learn. And because a pasture is a variable living community of plants, sampling must be done carefully or your results won't be accurate. Whatever tool you use to gauge dry matter per acre, it is important to take multiple measurements in each pasture, because any single sample may not be representative of the whole pasture. In addition, the amount of forage in the pasture changes, and during times of the year that growth is rapid the amount of available dry matter can change fast. To keep track of rapid changes, measurements may need to be taken frequently. For farmers new to grazing, it may be wise to measure the amount of dry matter per acre each time a new paddock is set up or the livestock are moved.

Whatever method of pasture dry matter measurement you choose, take time to learn how to do it right, and then practice. Find a local pasture group that meets regularly, or attend an on-farm pasture workshop to practice dry matter measuring and share ideas on grazing. There are also some online instructional videos listed in appendix E.

Hand Clipping

Hand clipping is the most accurate method to measure forage per acre. However, it is also slow, labor-intensive, and not very practical for most farmers. This is the method that was used to calibrate the other types of measuring devices, and we are very grateful someone has invented those devices to make our lives easier. For the sake of getting a better understanding of pasture forage measurements, let's briefly review how clipping is done.

To achieve a representative sample, several areas must be clipped, dried, and weighed in each pasture. The more variable the pasture is, the more samples need to be taken. The clipped area must be a known size, such as 1 square meter, and the plants should be clipped to the same height the livestock will graze to. The clipped material is then dried at a low temperature (don't cook it at a high temperature!). This can be done in an oven set at 100° to 120ºF. Once it is oven-dried, the sample is weighed and calculations are done to convert to pounds per acre.

Obviously this method is too time consuming to be practical for most farmers, so let's discuss some of the simpler methods.

Measuring Forage Height

Measuring forage height is a very simple way to estimate how much dry matter per acre is available. The taller the pasture, the greater the yield. However, inches of vertical pasture can't be precisely translated into pounds per acre of forage, because pasture density also affects yield.

Making a conversion from height to pounds of dry matter per acre varies according to the type of plants growing in a pasture and the density of growth. Reference tables offer data that assists with the height-to-pounds conversion. Such tables list pounds per inch per acre based on pasture density (percent vegetative cover) and species mix. This allows you to estimate how much dry matter the livestock are harvesting based on how many inches of pasture they consume. Table 13.1 was developed by Jim Gerrish, and it supplies dry matter yield in pounds per inch of pasture height. It's a useful tool for estimating dry matter availability per acre. Attending pasture walks or discussion groups is a great way to learn how to make these estimates more accurately.

Researchers have calibrated pasture clip and weight measurements with pasture heights in order to generate tables such as this one. A generic number of 200 pounds of dry matter per acre inch is sometimes used, too. However, this estimate may be more or less accurate depending on the plant density, canopy structure, and plant species mix.

Other tables rely on the estimated hay yield of a field to roughly estimate the amount of pasture forage that could be produced per grazing rotation or per year. Keep

Figure 13.1. These two pastures are similar in height, but the pasture on the right has more diversity and density, so it will produce more dry matter per acre.

Table 13.1. Pasture Dry Matter (expressed as pounds per acre inch)

Pasture Species	Stand Density		
	Fair 60–75% Cover	Good 75–90% Cover	Excellent > 90% Cover
Tall fescue + nitrogen fertilizer	250–350	350–450	450–550
Tall fescue + legumes	200–300	300–400	400–500
Smooth bromegrass + legumes	150–250	250–350	350–450
Orchard grass + legumes	100–200	200–300	300–400
Bluegrass + white clover	150–250	300–400	450–550

Source: Jim Gerrish, Management Intensive Grazing: The Grassroots of Grass Farming (Green Park Press, 2004).

in mind that poor management — particularly not allowing plants to fully rest and regrow after each grazing — will lower yields significantly. Poor soil fertility or plant species selection may also result in lower-than-standard yields.

USING A GRAZING STICK

Grazing sticks are the simplest and most practical tool for most farmers to measure pasture dry matter. At a glance, these look like just a ruler or yardstick. However, a grazing stick is actually a calibrated measurement system. With this low-tech device, you don't need to look at a separate table to figure out what the height measurements add up to in pounds. All the information is printed on the stick.

The grazing stick includes a ruler for measuring plant height, some grazing guidelines, a grid and directions for measuring plant density, and conversion formulas to turn the height and density numbers into pounds per acre. The grazing stick that is in use today is based on a Soil Conservation Service (now known as NRCS) tool that was greatly improved through research by Dr. Darrell Emmick. Emmick added many features to the stick, including a low-tech system to measure pasture plant density. Grazing sticks are made by many organizations in the United States and Canada, and may vary somewhat because they are calibrated to different regional forage species mixtures.

When you're using a grazing stick, remember that the measurement is only as good as the sample areas you measure. If the pasture has a lot of variation in height and density, you will need to take more samples to come up with an accurate representative value overall. There are some online instructional videos on how to use a grazing stick listed in appendix E, as well as information on where you can get one.

Step one: Use the ruler on the stick to measure total forage height. Directions on the stick will remind you to subtract the amount of residual plant height left behind after grazing. So the number of inches of height used in this calculation will be the total height minus 3 inches, which is the amount of residual most grazing sticks instruct you to subtract.

Step two: Measure the density. Slide the stick through the plants so that it's flat on the ground at the soil surface with the density meter (a simple grid with a series of dots scattered over it) faceup. Stand up and look directly down at the density meter. Count the number of dots you can see. Now you can look at the information printed on the stick and see how many pounds of dry matter there is per inch of height.

Step three: Multiply the number of inches from step one (total height minus 3 inches) by the pounds of dry matter per acre inch. Now you have an estimate of how

Figure 13.2. One feature of a grazing stick is a ruler for measuring plant height (left). The stick on the right has the density measurement grid faceup. You can use this face to assess the amount of ground surface covered by forage. The more dots on the grid visible through the pasture canopy, the lower the plant density.

many pounds per acre the livestock can harvest. And don't panic! Remember that all these instructions are also printed on the stick! Also remember that this stick is calibrated to give you the number of pounds available for the herd or flock to actually graze and harvest per acre.

To get an accurate measurement, take these measurements in multiple locations throughout the pasture. Plant height and density will usually vary in each pasture; by taking several measurements and averaging them, you will be less likely to over- or underestimate how much feed is in the pasture.

Using a Plate Meter

A plate meter measures pasture height and also factors in plant density. These devices are sometimes called rising plate meters or falling plate meters. The meter is a disk or plate that slides freely on a measuring stick inserted through a hole in the center. The plate descends onto

Figure 13.3. This rising plate meter factors in both height and density to give a measure of how much dry matter there is per acre. Photo courtesy of Susan Monahan, Northwest Crops and Soils Program, UVM Extension.

the forage, and the density of the pasture determines how far down it settles. This allows you to take a height measurement on the measuring stick while factoring in plant density. This improves the accuracy over a simple height measurement.

Plate meters are available in a range of plate weights and sizes, and each one needs to be correctly calibrated. For example, the model that Iowa State University makes is a measuring stick and a 50 x 50 cm square of Plexiglas that weights 2.6 pounds. This is calibrated to represent 263 pounds of dry matter per acre per inch on the measuring stick.[1] The cost of plate meters also varies widely; there are even some very simple versions that you can make on the farm. However, if you make one yourself, it is important to take the time to make sure it is giving you accurate measurements.

In addition to the rising plate meter there is another, often quite costly, pasture dry matter measuring device usually called an electronic capacitance meter. This is a computerized, handheld meter that directly measures total dry matter in a pasture. Due to the cost, these are generally used by researchers; not many farmers own them.

Figuring Daily Acreage Requirement

Once you have figured out how much dry matter a pasture can supply, you can calculate the number of acres needed to feed the herd or flock for one day. To do this, divide the amount of dry matter required by the group of animals per day (dry matter demand) by the amount of dry matter available per acre.

required DM per day ÷ available DM per acre
= acres required/day

Let's look at an example. Our sample farm has a herd of 50 dairy cows that graze on pasture and are also fed some grain and dry hay in the barn. The portion of their dry matter demand (DMD) from pasture is 30 pounds DM/cow/day; the rest of their feed intake requirement is provided by grain and dry hay. You can figure out the total daily DMD from pasture for the entire herd:

30 lbs/cow/day x 50 cows = 1,500 lbs/day

Let's assume the farm has an estimated available DM of 1,200 pounds per acre. To calculate the daily acreage requirement, divide the DMD by the available DM per acre:

$$1,500 \text{ lbs/day} \div 1,200 \text{ lbs/acre} = 1.25 \text{ acres/day}$$

This just gives us the number of acres of pasture to feed this herd for one day. It also assumes that animals are being turned into well-managed pasture daily — one with a density and height that mean it really is able to provide 1,200 pounds DM per acre per day. This calculation won't work for pastures that are continuously grazed for long periods of occupation or are poor quality.

Once you know how many acres are required per day, you can start to calculate how many total acres will be needed over the whole grazing season. Some additional information will be needed first, including the planned period of occupation, what stocking density you want to use, and the expected pasture regrowth period. Let's discuss stocking density next.

Stocking Density

Stocking density is the number of animals, or pounds of animals, per acre at a given point in time. Stock density is not the same as stocking rate, which is the total number of animals on the whole farm. A farmer increases stocking density by putting more animals into a smaller area. So adding more animals or making smaller paddocks creates higher stock density. The smaller the paddock, and the more often you are willing to move the herd to the next paddock, the higher the stocking density can be.

A higher stocking density has several advantages over low-stock-density grazing, as described in previous chapters. Higher stock density reduces selective grazing, creates a more even manure distribution, and stimulates animals to eat more when they are moved more frequently. When managed correctly, high-stock-density grazing will create a pasture with higher plant density. In addition, because animals are moved frequently, they have access to a more consistent feed quality and quantity. But there are some challenges with using higher

stocking densities. Not all farmers will find grazing with a high stocking density possible or practical.

Overall, stocking density decisions are based on two primary factors:

- Paddock size — how often the farmer is able to move the herd or flock.
- How much forage is available per acre.

Let's look at a specific example of how a farmer can play with stock density. In this example, 50 Jersey cows weighing 1,000 pounds each are given a fresh paddock 1.25 acres in size once a day.

$$50 \text{ cows x } 1,000 \text{ lbs/cow} = 50,000 \text{ lbs}$$
$$50,000 \text{ lbs} \div 1.25 \text{ acres} = 40,000 \text{ lbs/acre}$$

The stock density in the paddock is 40,000 pounds per acre.

If the farmer decided to divide the paddocks in half (resulting in paddocks of 0.625 acre) so the cows can have fresh pasture twice a day, the stock density would double.

$$50 \text{ cows x } 1,000 \text{ lbs/cow} = 50,000 \text{ lbs}$$
$$50,000 \text{ lbs} \div 0.625 \text{ acre} = 80,000 \text{ lbs/acre}$$

Many dairy farmers provide the lactating herd with a fresh paddock after each milking, so this is a reasonable stocking density for that type of management system. Some farms graze with higher or lower stocking densities than this, depending on their farm goals and how often they are able to move the herd. If there is more dry matter available per acre, or if farmers are able to move the herd more frequently, they can attain an even higher density.

However, it isn't always practical to manage for a high stock density, because it requires more intensive involvement by the farmer. A farmer who has an off-farm job may only have time to move the herd to a new paddock twice a week. This is still frequent enough for good results, but it will need to be with larger paddocks and therefore a lower stocking density (otherwise the animals could run out of food). In this scenario, pastures will improve more slowly, and feed intake and quality will be more variable for the herd. However, for a farmer with a busy off-farm work schedule, this system can work well for many types of livestock such as a beef herd or dairy heifers.

For example, 40 beef cows with 30 calves have an estimated total weight of 52,000 pounds during the spring. If they are moved to a new paddock of 4.25 acres every three days, the stock density would be just over 12,200 pounds per acre. Compared with the dairy cow scenario on page 162, this is a lower stock density. However, at this density, moving the herd every three days and giving the pastures ample time to regrow after each grazing, these pastures will still improve over time.

Let's say the farmer quits her off-farm job and changes her management approach. She decides to move the herd once a day into paddocks that are 1.4 acres each. The stock density will increase to just over 37,000 pounds per acre. If she decides to try a much higher stock density by using strip grazing, moving the animals four times a day with paddock size of 0.35 acre, the stock density will be over 148,000 pounds per acre. Note that in all three scenarios, the farmer is working with the same number of animals and the same total amount of pasture area. What is changing is the density of animals in the pasture and the portion of total pasture that the animals occupy at any point in time.

So we can see that figuring out the appropriate stock density for a farm depends on how often it is practical to move the animals. But it also depends on how much forage a pasture has to offer. If the amount of dry matter available to graze per acre is low, the stock density has to be proportionally lower than when dry matter per acre is high. Pasture plant density is discussed in chapters 3 and 5, and strategies to maximize pasture dry matter intake by livestock are discussed in chapter 9.

Paddock Power

More paddocks provide the farmer with "paddock power." The more paddocks there are, the shorter the amount of time animals spend grazing each one, so they are less likely to graze plants too short. Perhaps even more important, the more paddocks there are, the sooner the pasture plants get to start resting, recovering, and regrowing after being grazed. More paddocks also allow grazing with a higher stock density.

A pasture divided into two paddocks is better than one paddock, since at least half the plants in the pasture are getting a chance to regrow at all times. But if those two paddocks are each divided in half, there are now four paddocks and three-quarters of the plants are regrowing while one-quarter is being grazed. The more times the paddocks are subdivided, the more paddock power there is. More paddocks gives the farmer the power to more rapidly improve pasture quality and provide a higher quality and more consistent feed to the animals.

Adding more fixed paddocks built with permanent fence is not the only way to achieve higher stock density. Techniques to increase stock density can also include strip grazing using portable posts or tumble wheels. Automatic gate openers or Batt-Latches are also helpful. It is important to move a back fence behind the herd at least every few days, however, to prevent animals from regrazing areas they have already grazed. For more detail about using fencing to increase the number of paddocks, see chapter 14.

Figuring Total Acreage Needs

Now we have calculated how many acres are needed per day, or the daily paddock size and how that relates to the stocking density. Next, we can add in information on the length of the regrowth or recovery periods and calculate the total number of acres of pasture needed. To put this simply, you multiply the length of the required recovery period times the daily paddock size.

regrowth period x acres required per day
= total acres needed

We actually need to add one more paddock to this calculation, since the paddock the animals are in can't start its regrowth period while they're in it! Lets return to the 50-cow dairy farm example presented earlier in this chapter: When the pastures need 24 days to regrow, 31.25 acres will be needed (24 days x 1.25 acres per day = 30 acres plus one more 1.25-acre paddock).

In this part of the calculations, it is important to remember that recovery periods vary in length! As the length of these required regrowth periods varies, the

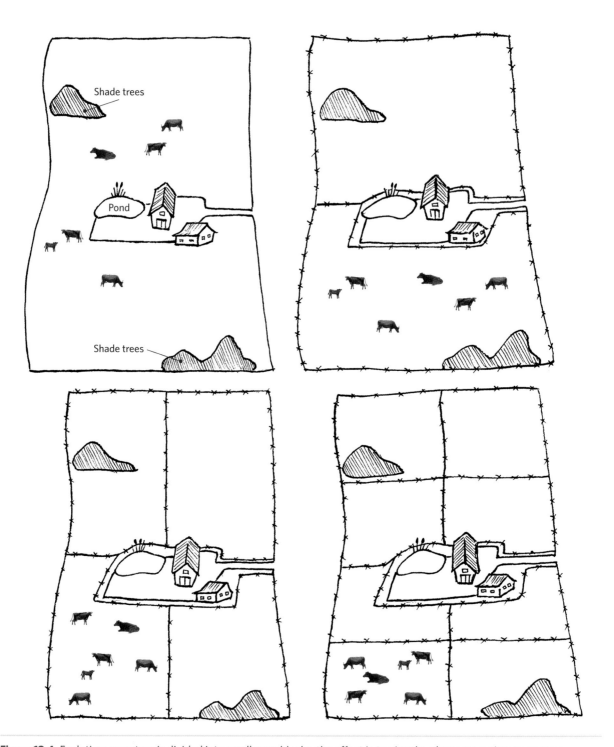

Figure 13.4. Each time a pasture is divided into smaller paddocks, the effect is to give the plants more time to regrow between grazings. In addition, stock density can be increased. With longer regrowth periods and higher stock density, the farmer has the power to more rapidly improve pasture quality. Subdividing into smaller paddocks requires not only additional fence, but planning to provide water in each paddock and access to shade when necessary. Illustration by Anna Powell and Marcia Brewster.

number of acres needed will also change over the growing season. To avoid overstocking the farm, creating overgrazing damage, start by determining the number of acres needed at the time of year when the recovery periods are the longest. This quickly tells you what a sustainable stocking rate is for the pasture or farm. This will help with the planning process to make sure the farm is not overstocked.

Matching the flock or herd size to the land base using maximum recovery periods makes it easier to avoid damage to pastures. Farms with too many animals and not enough land will find it very challenging to avoid overgrazing damage to the pastures at some times of the year. So if the maximum recovery period on our sample farm is 40 days, and the herd needs 1.25 acres per day, here are the calculations to figure the total acreage, factoring in one extra paddock from which the flock or herd will begin their rotation:

$$40 \text{ days} \times 1.25 \text{ acres/day} = 50 \text{ acres}$$
$$50 \text{ acres} + 1 \text{ extra paddock } (1.25 \text{ acres}) = 51.25 \text{ acres}$$

This calculation is for the time of year when regrowth is slowest, which is mid- to late summer in most areas. Thus, in practice some of that 51.25 acres will need to be hayed, clipped, or grazed with other livestock earlier in the year when growth is faster in order to maintain dairy-quality pasture forage. So in spring, 31.25 acres will be grazed while 20 additional acres are harvested for hay. Later, when growth rates slow, those 20 acres are added into the grazing rotation and a total of over 50 acres is grazed. Because growth rates vary, if you want to accurately plan the acreage needs throughout the grazing season, you will want to work this calculation for several different times of the growing season.

The worksheet in appendix B, on page 208, takes you through the calculations for figuring acreage requirements step by step. First you calculate acres needed per day, or daily paddock size; then you calculate the number of acres needed at different times during the grazing season. This allows acreage to be recalculated each time the length of the regrowth period changes.

Using this worksheet is just one of several methods of doing the pasture planning math. Whatever method you use, though, be sure to plan for variable recovery periods. It is also essential to be prepared to replan, because in any given year plants may grow faster or slower than you predicted, there may be more or less dry matter actually available per acre, or the herd size may change.

Several other useful methods of calculating and planning stock densities, paddock sizes, and acreage needs can be found in *Management Intensive Grazing* by Jim Gerrish and *Animal Production Systems for Pasture-Based Livestock Production* edited by Edward Rayburn. Planning can be done by hand on worksheets or directly on a copy of the farm map. There are also apps and online planning resources that can help with the planning and mapping by recording and even calculating how much dry matter the herd gets from pasture. Holistic Planned Grazing includes a very comprehensive planning process, which includes either paper planning worksheets or computer planning spreadsheets. See appendix E for more information.

Plan for Recovery Periods

Designing a grazing system that allows you to manage variations in how long the plants really need to fully recover is essential. If you don't plan for variable recovery periods, it will be impossible to prevent overgrazing damage from rotating back into paddocks that have not fully recovered. So having the right number of paddocks and/or acres is very important to build into the grazing plan. Although we have already calculated the paddock sizes and number of acres needed to graze throughout the growing and grazing season, let's look at this slightly differently now to make sure we plan for both the right number of paddocks and the right number of acres.

Let's return here to some of the fundamental principles of good grazing management — the interrelationship between the grazing period and the recovery period. This has an effect on the number of paddocks needed. As we know, the grazing period, or period of occupation, is the number of days the herd or flock is left in a paddock before moving to the next one. The recovery period is the length of time before that paddock is ready to be grazed again. If there are not enough paddocks, and the herd is moved from one paddock to another too fast, it will compromise the recovery

Table 13.2. How Many Paddocks Are Needed?

Frequency of Moving the Herd	Length of Regrowth Period	
	21 days	30 days
Every 3 days	8 paddocks	11 paddocks
Once a day	22 paddocks	31 paddocks
Twice a day	43 paddocks	61 paddocks

period. The animals will end up grazing plants that are not ready to be grazed again. This will cause both the livestock and plants to do poorly.

Because the length of the recovery period varies throughout the year, either more paddocks must be added to the rotation as growth slows *or* the number of days in each paddock must be increased (and supplemental feed may need to be added to the ration) to slow the rotation speed. However, leaving the animals in the paddock longer may not be good for the plants or animals because the animals may be forced to graze the pasture too short.

The best way to deal with variable recovery periods is to plan to have as many paddocks as needed to match the *slowest* recovery period on the farm. Some of those paddocks may need to be clipped or harvested for hay during times of the year when plant growth is rapid. But

by having enough paddocks, the common mistake of rotating a herd too fast can be avoided.

The formula to calculate the length of the period of occupation is:

recovery period ÷ (total number of paddocks − 1) = grazing period

For example, if a farm has 11 paddocks, and 30 days are required for the paddocks to regrow fully, the herd must be left in each paddock for a 3-day grazing period to prevent them from being rotated too fast.

30 days ÷ (11 paddocks − 1) = 3 days/paddock

If the farm wants to move the herd to a new paddock once a day, or twice a day, a lot more paddocks are needed to allow each one time to regrow after grazing, as shown in table 13.2.

Generally at least 8 to 10 paddocks are needed for good grazing management. With this number of paddocks, the grazing period can generally be kept to four days or less. With a grazing period of four days maximum, it's much less likely that animals can regraze regrowing plants. It also lowers the chances of plants being grazed too short.

CHAPTER 14

Grazing System Design and Infrastructure

Designing a new grazing system, or redesigning an existing one, involves decision making on land usage, paddock sizes and shapes, lanes, fences, water systems, and how to fit it all into the landscape. A poorly designed grazing system will be frustrating at best. At worst it won't be practical to work with and can result in damage to the pastures, along with poor livestock performance. A well-designed system needs to be flexible, practical, and carefully fitted to the uniqueness of the farm's landscape.

The importance of having well-designed, functional electric fence cannot be overstated. Many farms have set up grazing systems with paddock subdivisions and high hopes of rotating the animals through the paddocks, only to find that the animals don't respect the fence. It only takes a few escaping animals to knock down the fences and ruin the well-planned grazing rotation. This is usually due to an undersized energizer, or to poor installation with an undersized ground. Information on how to make sure the fencing actually works is included later in this chapter.

It is also important to train new or young animals to electric fence before putting them out in the pasture. This is most effectively and safely done in a training yard where there is an exterior visual barrier or fence such as a rail fence. Inside that fence an electric fence (with high voltage!) is set up so that as the animals touch the fence and receive a shock, they see the visually obvious fence behind the electric wire and back up instead of running straight through the electric wire (see Training Paddocks and Sacrifice Pastures later in this chapter).

Taking Inventory

To start, it is important to have a good inventory of the land base and other resources. This can be done by walking the farm to look at soils, slope, aspect, drainage, and vegetation. Use an accurate map that provides detail on the available acreage and soil types. Don't just trust your assumptions about how many acres of pasture you have. The actual amount of open useful land may be a very different number!

Farm acreage maps are available from your local FSA and NRCS offices. While at the NRCS office, inquire about programs that may be available to help with the costs of building and managing a grazing system. They may also be able to provide soil maps and other resources. If maps from NRCS are not available, there are apps and websites that provide maps that show acreage of individual fields and pastures. A good map will show the acres of each field as well as the locations of streams, forested areas, hills, and other features. Maps which show the soil types are now available online for many regions. Refer to appendix E for more details on online resources.

Using the map along with time spent hiking around the land, take some notes on the land base inventory:

- How many acres of rough, hilly pasture are there?
- How many acres of tillable high-quality cropland are there?
- How many acres can be both grazed and mechanically harvested or hayed?
- Where are the swamps, wetlands, and riparian areas?

- Where are there natural water sources that might be used to supply drinking water?
- Are there water features, wet areas, or other parts of the farm that livestock need to be fenced out of some or all of the time?
- Where are there areas with shade?
- How many acres of droughty soils are there? How many acres of seasonally wet soils or areas with poor drainage? How many acres of prime agricultural soils?

You may also need to consult the soils map in order to answer some of these questions. (See chapter 4 for more about soil maps.)

Subdividing the Land

Using either the online mapping program or colored pens and a paper copy of the map (and some common sense), you can decide where to put lanes, fencing,

paddocks, and water. Some farmers find it helpful to have large photocopies of the farm map laminated; they then use dry-erase markers to mark pasture subdivisions and infrastructure. Once the grazing system is designed, these laminated maps can be kept on the wall of the barn or office for future planning and even grazing record keeping.

Pasture or Hay Land?

It can be helpful to start by dividing the land available into two categories. First list the land and acreage to be used as permanent pasture. Then list the land and acreage that will be mechanically harvested, but may also be grazed. It makes sense to use more permanent fence in areas that will be permanent pasture. In areas that will also be hayed, it may still be helpful to have a permanent perimeter fence, but interior subdivision fences can all be done with portable fence to make it easier to cut hay without obstructions.

Figure 14.1. Using a permanent electric fence around the exterior of the field and portable fence for interior subdivisions offers more flexibility to take a cut of hay from the field, spread manure, and adjust paddock sizes. On this farm, the paddock size can be adjusted using polywire and tumble wheels so the stocking density is ideal for maximizing pasture dry matter to the herd and managing the pasture plants. Photo courtesy of Eric Noel, Health Hero Farm, Vermont.

Paddock Sizes and Shapes

Once you know the number of pasture acres per day needed to feed the herd or flock, you can make decisions about where to set up the pastures. It can be helpful to create large paddocks — also called grazing cells — to provide several days' grazing to the herd or flock. Those areas can then be subdivided as needed with portable fencing.

When subdividing, particularly on a new farm, it is helpful to have fewer permanent fences. Instead, focus on putting the permanent fences along the large field perimeters. This will keep the system as flexible as possible to allow fences and water lines to be moved as the herd size changes and pasture quality improves.

Remember that the number of acres of pasture needed will change during the season as the speed at which the pasture plants are growing changes. If possible, a grazing plan should include some flexibility so that some land can be mowed or harvested during part of the season and grazed at other times.

The closer to square a paddock is (rather than long and narrow), the better the animals will graze it. In addition, it usually takes less fence to create a square paddock than wedges or narrow strips. However, unless the farm is perfectly flat and uniform, the actual shape of the paddocks will depend on natural landscape features or past land use.

Permanent Subdivisions

Whenever possible, permanent fences should divide the farm so that areas that grow very differently from one another can be in separate paddocks. For example, if one area in a field is droughty and slow growing, it would be easier to manage fenced separately from a fast-growing area in that same field. If one paddock has both areas that grow rapidly and others that grow slowly, it can be challenging to determine when the plants are all fully recovered and ready to be grazed again. By the time some plants are ready to graze, others will be overmature and less nutritious.

For farmers who have been managing the farm for several years or longer, it's easy to know where the different-growth-rate areas are. On a new farm, however, it can take some time to learn how each area will grow at different times of the year. This is one good reason to put in fewer permanent fences in the first year, using as much portable fence as possible and taking some time to decide where it makes the most sense to put the permanent subdivisions.

Logical places to put permanent subdivisions include contour lines, where a field changes from convex to concave, between south- and north-facing slopes, along dividing lines between very different plant communities, and along the edges of forested areas or buffers to streams and rivers. It can also be helpful to consult a soil map to see if there are significant soil type changes. Plants growing in clay soil will respond very differently from plants in sandy soil, particularly in a wet year.

Some areas of a farm's landscape are concave and some areas are convex. In daily discussion these are called the low wet areas and the ridgetop dry areas. Depending on how wet the concave areas are, and how dry the ridges or convex areas get, these pastures may grow at very different speeds and produce different plant species. If the areas are large enough to be fenced separately, it will be much easier to manage them as separate pastures.

In addition to the shape of the land, the aspect or direction it faces can also influence how plants grow. A south-facing slope may warm up earlier in the spring, promoting faster growth, so it may make sense to fence that area separately and start grazing it earlier. Later in the summer when it is hot and dry, it will be convenient to have that area fenced separately so it can be given a longer rest period.

Lanes

For a dairy farm, lanes are necessary to prevent animals from walking through and damaging pastures during their daily travels back and forth from pasture to milking. Even for a beef or sheep flock that doesn't require daily access to the barn, using lanes when moving animals from one pasture to another has many advantages for the pasture plants. Using lanes prevents livestock from trampling and nibbling plants in pastures that still need more recovery and regrowth time. During wet spring and fall weather, having well-built lanes in the right locations will make managing pastures much easier. Dry, well-surfaced lanes can also prevent hoof injuries, keep animals out of the mud, and prevent erosion.

Figure 14.2. Lanes on this dairy farm have been built with a layer of road-building fabric under the gravel. The top layer of gravel is fine and round so it is comfortable for the cows to walk on, and it's maintained so water runs off to the sides. These lanes are built wide enough for two or three cows to walk side by side, and for farmers to drive an ATV on as they move water tubs, polywire, and posts.

A farm with lanes also has more options on where to move livestock, which can allow more careful grazing management. Different parts of the farm grow at different speeds at different times of the year. Although it is easy to get in the routine of moving animals from one paddock to an adjacent paddock, it's better management to move the herd to a paddock or pasture that is at exactly the right stage of regrowth. Having a lane to move them through makes it much easier to graze the paddocks when they are ready, instead of in order of proximity.

Lanes may be created simply by setting up a single strand of polywire parallel to the perimeter fence along the edge of a field. But they often need to be carefully sited and constructed to avoid mud and erosion and causing the animals sore feet and other problems. This is of particular importance on a dairy farm, where the herd needs to walk back and forth to the barn for milking several times a day. In wet areas or heavily used lane areas, a road-building filter fabric under gravel will create longer-lasting lanes that require much less maintenance. In some areas, additional work such as culverts and erosion control may be needed.

It is important to select the correct size and shape of stone for the top gravel layer of lanes. If a larger, sharper crushed stone is used, this can injure hooves; animals will avoid using the lane if possible.

Here are some suggestions for locating and building lanes:

- Site them on the contours or tops of ridges whenever possible.
- If possible, use them as one of the subdivisions of the pasture. Running the lane through the center of the pasture allows for paddocks to be closer to square rather than long and narrow.
- Avoid wet areas when possible. If the lane must cross a wet area, install a culvert or use road-building fabric under a layer of gravel to create a high, dry area.
- When lanes have to cross a stream, it is important to build them to prevent bank erosion or water-quality issues. NRCS can be a helpful resource in designing stream crossings.
- When possible, avoid lanes on steep slopes. If this is unavoidable, add waterbars to prevent erosion.
- Make the gate to the paddocks the same width as the lane. That way the gate can be used to block the lane or the paddock entrance to make livestock movement easier.
- Set up a gate at the corner of the paddock that is closest to the barn so that animals move easily out of the paddock.
- Locate gates from the lane into the paddock in areas that won't get muddy.

Talk to the local NRCS or conservation district office to see what programs are available to help pay for engineering or construction of well-designed lanes.

Training Paddocks and Sacrifice Pastures

Having an area where livestock can be moved and fed stored feed during bad weather, floods, or droughts is important. If a secure permanent fence is built around this heavy-use area, it becomes a location to which animals can be moved in order to keep them safe and prevent damage to pastures. It is particularly important to keep animals off pastures when soils are saturated and/or freezing and thawing to prevent soil compaction and damage to the plants.

This area can also be used to train young animals or new animals to electric fence. The ideal way to do this is to build a fence that provides a visual and physical barrier, and then have an electric fence just inside the barrier fence. Livestock will then naturally back up instead of running through the fence when they are shocked for the first time. It is very important, and actually safer, to maintain high voltage on the fence during these training periods. Animals unfamiliar with electric fence are more likely to mess with fence that has low voltage and may get tangled and injured or killed. A single shock from a well-installed low-impedance energizer will quickly teach them to safely respect the fence!

SHADE

When managing livestock during hot weather, it is important to consider temperature, humidity, the type and quantity of shade available, the types of facilities on the farm, cattle breeds, and how well the livestock are adapted to the local climate. In some areas it will make sense to design a significant amount of shade into some of the pastures by using hedgerows, forest edges, or ideally spaced trees. See figure 4.9 for an example of a tree-shaded pasture. Figure 14.4 is a nice example of a well-planned lane on a dairy farm, where trees are used to shade the cows as they walk back and forth to the pastures.

In more temperate climates where there are few hot, humid days that might create heat stress, shade areas may not be needed as often. During cool weather,

Figure 14.3. On a cool spring day, these sheep still habitually crowd up under their shade tree even though it isn't hot enough for them to need the shade. This behavior results in excess manure and trampling under the tree.

Figure 14.4. At Spring Wood Organic Farm in Pennsylvania, many of the lanes are lined with shade trees so the dairy herd has an easier time walking back and forth to the pastures in the heat of summer. When it gets very hot, cows will walk back to the barnyard, where they eat balage and stand under cooling misters.

animals may choose to congregate in shady areas anyway. By doing this they are concentrating hoof impact and dropping most of their manure in one place instead of spreading it around the pasture evenly. In some cases this may make them hotter and increase problems with flies.

For most farms, the ideal is to have some specific shade pastures available to use on hot, humid days to prevent heat stress. For some dairy farms, the herd can be kept inside the barn during the hot days of summer, with fans or even a misting cooling system during the peak heat hours. They are then put out to pasture during the evening, night, and morning hours when temperatures and humidity drop.

A few farms have found ways to provide portable shade or portable misting systems to keep livestock cool on pasture. However, these are not always practical. Some are heavy and difficult to move, while lighter-weight versions may blow over on a windy day. Farms with a larger herd size or using high-stock-density grazing with frequent herd moves will also find portable shade systems less than practical. Interestingly, many farmers report that when their livestock are grazing high-quality pasture, they appear to experience less heat stress.

Infrastructure for Fences and Drinking Water

Programs to help with the design and costs of fence and water systems are often available through local conservation districts or the USDA Natural Resources Conservation Service. They can also help you make sure you are not violating any water-quality or required agricultural practice laws that may limit how close manure and livestock can be to streams, rivers, or other sensitive areas.

To start with, try to be as flexible as possible with your fence and water systems. You may change your setup a few times, and you may need to have flexible paddock sizes as your herd size changes and your pasture productivity increases.

ELECTRIC FENCE

Electric fence technology makes good grazing management much easier than older stone walls, rail fences, and barbed wire. However, there are places where a physical-barrier-type fence may also be useful. In the training paddock, barnyard, or heavy-use area it may make sense to have a physical-barrier fence instead of just electric fence. Out in the pastures, however, most of the time electric fence will be the lowest-cost and easiest-maintenance option, and allow for the best and most flexible management.

If you have no experience with electric fencing, there are some resources listed in appendix E that provide additional information. In addition, many local fence installers teach an annual fencing workshop for beginners. Such workshops are a great opportunity to gain some hands-on experience.

Because electric fence is a psychological rather than a physical barrier, using it successfully depends on having livestock that have already learned that touching the fence is painful. A good-quality energizer that is well grounded and installed correctly is essential. A well-built fence with many wires will not restrain the livestock if the energizer is too small or poorly installed. But a single strand of electric fence, if built well and charged correctly, can keep in a whole herd of cattle. For sheep and goats, multiple strands will be needed, and many small ruminant farms use electric net fence.

The Energizer

Start with a good-quality, low-impedance energizer and ground it correctly. Most have some information printed on the energizer or box to tell you how powerful they are. This may tell you how many miles of fence they can run, or how many joules. But don't rely on what information is printed on the packaging. It is best to buy the energizer from a reputable dealer who can help make sure you purchase the right unit. For example, a sheep farm with multiple-strand fencing may need a larger energizer than a beef farm that plans to use just one strand of wire. An experienced local dealer will also provide repair service as needed and other useful materials such as lightning protection and coated high-tensile lead-out wires.

In addition to size, energizers also vary with the power source they use, and can include 110-volt, 220-volt, batteries, or solar panels. Battery-powered energizers can be very problematic due to the battery power running out when it is needed most. So unless it is a truly remote location, it is always better to try to run a fence wire from a plug-in energizer to the field instead of using a battery/solar energizer. Energizers that run off 110 or 220 electricity will provide a lot more power per dollar spent and will work best in the long run.

Make sure the wires carrying the voltage to the fence and the wire connecting the energizer to the grounding system are not too small (don't use house wire!). It is also important that all the connections are good. There are good-quality 12.5-gauge coated lead-out wires that are ideal for carrying the high voltage from the energizer to the fence. Once the energizer is installed, test the ground to make sure it's large enough.

Figure 14.5. This high-quality energizer has an output value of 36 joules and is powering many miles of fence on this dairy farm.

A good digital fence tester is a worthwhile investment. Modern testers detect the voltage on your fence, and can even tell you if you have a short and how much power you are losing to that short. They even use an arrow on the readout to tell you which direction the short is in to make it easier to find it and fix it.

Some of the more expensive modern energizers have remote shutoff options. This allows you to be out working on finding the short while the energizer is on. Then once the problem is found, you can use the remote to shut off the energizer from anywhere on the farm. Once the fence is repaired, just turn the energizer back on remotely.

Grounding

The most common cause of an electric fence system not working well is undersized grounding systems. In order for the animals to get a shock from the fence, the electric circuit needs to be completed. This means the charge must pass from the fence to the animal to the soil and back to the energizer by way of the grounding system. If the grounding system is too small, the animal will only get a weak shock even though the energizer might be large. This results in escaping livestock, frustrated farmers, and damage to the pastures as livestock graze wherever they want!

A simple way to know if the grounding system is adequate is to short out the live fence to a dead short. This can be done by connecting a metal gate, or metal rods that have good contact with the soil, onto the fence. Obviously this is best done when the energizer is turned off! Once the short is created, turn the energizer back on and test the fence. For the ground test to work, it's best to get the fence voltage down to 2,000 or fewer volts, then use that same digital fence tester to test the voltage on the ground wire. If the test shows significant voltage (over about 300 volts) on the ground wire, the ground system is not adequate.

If the ground needs to be made bigger, don't add the next ground rod close to the existing ones. Go 50 to 100 feet before putting in another rod. The more nonconductive the soils in the area are, the farther apart these rods need to be. Some farms with very nonconductive gravel soils have found they need grounding systems significantly larger than what is described in the instructions that come with the energizer!

The grounding system must be located away from any metal plumbing, water tanks, or electric power grounding systems. This will help prevent problems with the energizer becoming a source of stray voltage.

Wire and Fence Posts

In addition to using a highly conductive lead-out wire, it is also helpful to have high-quality perimeter fencing that can conduct electricity with minimal resistance. High-tensile fence using 12.5-gauge wire requires the least maintenance and has the best conductivity over time. However, other types of steel wire fence of at

Figure 14.6. For animals to get a shock from the fence, the electric circuit needs to be completed. This means the charge must pass from the fence to the animal (or the farmer in this illustration!) to the soil and back to the energizer by way of the grounding system. Illustration by Anna Powell.

least 14 gauge can also make a good perimeter fence, though they will usually require more maintenance over time.

Posts may be made of treated wood, natural wood, fiberglass, metal, or plastic. Be aware that most treated wood posts are prohibited in new installations on certified organic farms (see appendix D for more information on organic certification). Metal posts can be problematic in electric fencing because if the hot wire shorts to the post, it can drain a lot of voltage from the fence.

Various types of insulators are designed for use on wood posts, metal posts, portable posts, and even live trees. Some are insulating tubes that must be slid onto the wire during construction, while others are attached to the post and the wire snapped into the insulator later.

High-tensile fence requires bracing at corners and gates. Choosing which techniques and materials to use when building bracing depends on local soil conditions as well as what materials are locally available. Installing more corners and gates increases the cost and labor of fence building, so designing straight-line fences or gradual curves will cost less.

If you want to build your own fence, it's a good idea to attend a workshop on fence-building techniques. Many local fence dealers offer such workshops. You can learn how to tie knots in high-tensile fence, what types of braces, wire, and tightening devices work best, and what local fence posts are most cost-effective and available.

High-tensile wire is heavy, and can be awkward to install without the right tools. You'll need a good-quality set of pliers to cut the wire and tie knots. Wearing eye protection is also important! In addition, a spinning jenny will make it possible to unroll wire without tangles.

Once the perimeters of the fields are fenced, one of several types of portable fencing can be used to subdivide larger areas. These include highly conductive stranded or braided aluminum wire and many types of poly rope, polywire, and poly tape. The stranded aluminum wire is the heaviest to work with, but also the most conductive of the portable types of wires. There are different types of poly wires, which have more or fewer strands of conductive metal in them. Polywire that has 10 strands of metal in it will carry voltage for longer distances than polywire with only 3 strands.

It pays to do some research before purchasing energizers, portable posts, and polywire, because there are significant differences in quality. Some polywires are made using poor-quality plastic that breaks down rapidly in sunlight, along with few strands of metal to conduct electricity. Others include more metal and sturdier plastic

Figure 14.7. Unrolling, tying knots, and handling the wire of high-tensile fencing requires some specialized knowledge and equipment. This is one of several types of tighteners that can be used on 12.5-gauge wire.

Figure 14.8. This insulator is slid onto the wire during construction and stapled to the post later.

Figure 14.9. This high-tensile fence corner brace also serves as a gatepost where the spring-gate handles hook to insulators on the post.

Figure 14.10. This permanent fence is being built using plastic posts and a combination of polywire, poly rope, and high-tensile wire. Previously brushy, the pasture is being cleared and fenced so that it can be improved.

that will be more conductive and last for many years. In addition to polywire, thicker poly ropes and poly tapes with better visibility are also available, as is electric net fence. The latter is a net or mesh made of polywire, and most come with the posts built into the fence. These come in different heights and with different sizes of mesh openings for use with various types of livestock.

In addition to lightweight fiberglass and plastic posts, some farms also use tumble wheels. This type of "post" is designed for farms that do strip grazing on relatively flat fields. They offer advantages compared with portable posts, which need to be individually moved. The tumble wheel "tumbles" or rolls along when the polywire reel and polywire are moved. These are designed so the upper legs of the wheel are electrified while the lower legs — in contact with the ground — are not. This assures that the livestock don't push on them to move the fence by themselves.

Figure 14.11. This pasture is surrounded by high-tensile perimeter fence, with several paddocks set up using polywire on portable posts. The water tub is moved from paddock to paddock as the herd is rotated. Low-cost reels sold for storing extension cords are useful for winding up polywire.

Figure 14.12. This farm uses high-quality polywire on portable posts for interior paddock subdivisions. The portable water tub is filled from pipes connected to hydrants that have been strategically placed around the farm.

Figure 14.13. Portable electric net fence is ideal for small ruminants. High voltage is essential to prevent animals from challenging the fence and becoming tangled in it.

Figure 14.14. Electric net fence requires more work to move than a single strand of polywire. It also must be rolled up and unrolled correctly to prevent tangles!

Figure 14.15. Another way to manage a single strand of polywire is by using tumble wheels. As they roll or tumble, the two legs touching the ground are not electrified, but the upper ones do carry a charge.

Gates

Gates can be as simple as a piece of wire or polywire with a gate handle on one end, or as complicated as a Slinky-style spring gate connected to an automated gate opener such as a Batt-Latch release timer. When gates are part of a multipaddock system connected to a lane, it is ideal to have the width of the gate be the same as the width of the lane. That way the gate can be used to close off the lane when you're moving a group of animals. An alternative to this is to use a stretchy gate material such as a spring gate or a bungee poly gate.

Lightning Protection

Lightning or power surge damage to an energizer can happen in two ways. Lightning can hit the fence and a surge can travel back on the fence wire to the energizer. Or a power surge can come from the electric outlet.

Overall, the best form of protection from lightning is to make sure that the energizer is correctly grounded. In addition, there are also several types of lightning protection available for energizers. Since a significant amount of the energy surge that damages energizers comes from the electricity supply, make sure the energizer is plugged into a surge suppressor instead of directly into the power outlet.

The remaining risk of damage comes from power surges from the fence. Some of the tools to prevent this damage (other than a correctly installed large grounding system) include diverters, arrestors, chokes, and coils. Some of these devices, such as the arrestor or diverter, require their own grounding system. Others, like the choke or coil, don't need an additional ground.

TROUBLESHOOTING FENCE PROBLEMS

Once the electric fence is installed and running, it is important to regularly check the voltage and watch the animals to make sure everything is running correctly. Don't wait until the cows get out into your newly seeded field to notice that the fence voltage has been low for a week!

- Check the fence voltage at the lead-out wire *and* at the far end of the farm regularly. This shows if there are some shorts on the fence, or poor connections in the wires.

OPEN SESAME!

Batt-Latches or other types of automatic gate release timers can be set up to open the paddock gate at milking time to allow the cows to walk back to the barn or parlor. These can also be used to do high-stock-density grazing by moving the animals more times each day without having to hike out to the pasture to open another gate or move a front fence.

Using a Batt-Latch can allow a farm more time to work on other projects without having to interrupt the work every few hours to move fencing.

Figure 14.16. The Batt-Latch automatically opens this spring gate to allow cows on Elam Stolsfoos's farm in Pennsylvania to enter a new paddock after a few hours of grazing.

- Check the ground system regularly to make sure it's still big enough. Over time the ground can corrode or break and may need repair or more ground rods.
- When building fences, put in switches so that sections of fence not in use can be shut off. These will also make it easier to pinpoint where shorts are when troubleshooting.
- Create a system to make sure someone doesn't forget to turn the energizer back on or shut the gate. One

farm with a creative approach to this keeps a large unattractive beaded necklace hanging on the energizer. Whoever unplugs the energizer has to wear the necklace until they go back and turn it on again.

WATER

An ideal water setup is to provide water in each paddock, and to have the herd as close to the water as possible. The farther animals have to walk to reach water, the less evenly they will graze the pasture, and the more unevenly distributed the manure will be. A long walk to water can also decrease how much water the animals drink and in some situations this can lower dry matter intake from the pasture. Most people recommend having water within 600 to 800 feet of the animals. However, the practical reality of each farm's landscape and herd size may make that challenging.

How Much Water?

The amount of water animals drink depends on several factors, including temperature, humidity, and water content of the feed. In addition, the amount of water consumed will vary with different types of animals and their stage of production or growth.

If the water will be in the paddock and close by, each animal will be able to drink as it wishes. If the water is farther away and the whole mob wants to walk as a group to the source, a much larger amount of water or higher flow rate will be needed. This requires either a larger tank or a much higher flow rate.

How to Provide Water

If you're setting up a new grazing system, a stationary water system where animals must walk to drink from a tub near the barn or near a spring may be your only option to start with. By creating a lane that leads to the tub, it is possible to reduce cattle access and damage to pastures they have already grazed. Later, it will be ideal to be able to provide water in — or at least close to — each paddock.

In addition to the herd wandering back over previously grazed pasture to get to water, another downside of fixed water tubs is that the area around each tub can

Table 14.1. Water Intake

Type of Livestock	Range of Water Intake per Day
Dairy cow	15–25 gallons
Beef cow	12–20 gallons
Sheep and goats	2–3 gallons

Source: Edward B. Rayburn. editor, *Forage Utilization for Pasture-Based Livestock Production* (Natural Resource, Agriculture, and Engineering Service, 2007).

become a muddy, trampled, heavy-use site that may require some repairs or reseeding. In addition, without piped water to pastures animals may have to walk farther to drink.

One temporary solution is to park a wagon that holds a large tank in the pasture. This tank can then be used to either manually fill a tub or automatically fill a tub with a float valve. There are also several creative systems that can pump water from a pond, well, or stream to a tank.

If you're using a system to pipe water long distances to pastures, particularly when moving water uphill, there are several factors to consider to make sure enough water reaches the tub fast enough. The flow rate will start out based on the pressure and flow rate at the barn or well, but then that flow rate and pressure may drop due to traveling uphill, or because of resistance in the pipes.

Use of a larger-diameter pipe that offers less resistance will prevent the flow rate from dropping too much as it travels the length of the farm. When pumping water, the water pressure will drop by 1 psi (pound per square inch) for each 2.3 feet it goes uphill.[1] The flow rate and water pressure need to be high enough to fill the tub at least as fast as the herd or flock empties it while drinking. It is important to keep the tub full while the livestock are in the paddock. If the tub is empty, not only will the livestock be thirsty, but they may also roll the tub over and break the float valve.

Water lines can be buried, or pipe can be laid on the ground along a fence line or a lane. Many farmers prefer to keep the pipe aboveground so it is accessible for repairs and can be relocated if needed. The downside of aboveground water lines is that on warm sunny days, they may heat the water if they're not shaded. However, finding leaks in an aboveground system is much easier.

Figure 14.17. The float valve has a high flow rate, and the tub is connected to the aboveground water line with a length of high-quality garden hose. This float valve is connected to the bottom of the tub instead of the top, which makes it harder for curious heifers to break it (as long as the tank stays full!).

Figure 14.18. On his organic dairy farm in northern Vermont, Guy Choiniere shows a group of visiting farmers the portable posts, polywire, and movable water tub that make it easy for him to give the herd a fresh paddock with a full water tub after each milking.

Small portable tubs work well as long as the water flow rate is fast enough to keep them full while the herd drinks. Larger, permanent tubs are a better choice if the water flow rate is slower or if the herd is large. It is less expensive to buy a few portable tubs than to invest in many permanent tubs. Wherever the water source is, use a float valve to control water flow.

Drinking from ponds or streams may also be possible, but not in all locations, and you must pay careful attention to water quality and potential erosion. A strategically placed, well-designed access ramp to a natural water source can provide a secure, nonslip surface. This will prevent the animals from eroding the bank and standing in the water source, contaminating it for themselves and others downstream. Poorly managed access to ponds or streams can cause serious problems with both water quality and livestock health.

If it isn't possible to provide water using portable or fixed tubs in each paddock, then it is always best to use a lane for the animals to walk to the water instead of letting them walk through pastures that are starting to regrow. They will still be dropping some manure in the lane where it isn't needed, but at least they won't

Figure 14.19. Some farms prefer to use permanent drinking tanks, like this cement one with a built-in overflow pipe on Spring Pastures Farm in Maryland. These require careful design and management to prevent problems with mud around the base of the tub and in nearby areas.

be doing overgrazing damage to recovering plants as they walk.

Talk to the local NRCS or conservation district office to see what programs are available to help pay for engineering or construction of a well-designed water system.

This is also a good way to learn what regulations and restrictions there are if you're using streams or ponds as a water source.

Irrigated Pasture

Used correctly, an irrigation system can greatly improve pasture quality and productivity. But irrigation does add another layer of complexity to the grazing system design, paddock layout, and paddock rotation.

Pasture irrigation systems include flood irrigation systems, center pivot systems, traveling guns and irrigators, and pod line irrigation or K-line irrigation systems. Each requires a different amount of water, different water pressure levels, and very different design and setup.

Pod line or K-line irrigation systems can be used with low-pressure water systems and also work well on oddly shaped pastures. If a larger volume of water is available, a traveling irrigator or gun could be used. Center pivot systems are another option for farms with the right topography and sufficient water supply.

The first step in installing a new system is to find out what volume of water is available. This will narrow the options down to systems that will work with the water supply. Next a full cost estimate should be done of each system, along with some research on which system will work with the shape and topography of the fields and the plant species being grazed.

More information on irrigation system selection, design, and installation is available from local NRCS offices, and there are some resources listed in appendix E. In some regions, NRCS may also have programs to help farmers with the cost of irrigation. It's also important to make sure that the water source being used for irrigation is not restricted by state or local regulations, and it's essential that irrigation does not put local water or soil resources at risk!

Making the grazing rotation work with a new or existing irrigation system isn't always easy. There are some experienced graziers and consultants who can offer practical advice and ideas on how to use permanent and portable fencing along with irrigation to create productive and low-labor grazing systems. Jim Gerrish offers grazing schools that include hands-on practice setting up

Figure 14.20. This traveling irrigation system at Stonypond Farm in northern Vermont is supplied by a large pump that is able to supply a larger volume of water to the pastures than a pod line system.

and managing grazing systems with irrigation. Figures 14.20, 14.21, and 14.22 show examples of pasture irrigation systems. See Grazing a Large Dairy Herd at the end of chapter 15 for another description of integrating grazing and irrigation.

Managing Animal Flow on Dairy Farms

Designing a grazing system for dairy animals requires planning for efficient animal flow. The lactating flock or herd needs to be able to walk back and forth to the milking facility several times a day. This requires more planning for lanes and fence locations than on a beef farm, where the herd is simply moved from one paddock to another and may not return to the barn during the entire grazing season.

Lanes on a dairy farm need to be wide enough to accommodate the whole herd or flock, and they need to be well surfaced to keep the animals from getting muddy

INNOVATIVE IRRIGATION

Jim Gerrish is the owner of American Grazing Lands Services and the author of several books on grazing. He was previously a grazing specialist with the Forage Systems Research Center at the University of Missouri, and now runs a grazing operation in central Idaho. The ranch includes both dry rangeland and irrigated land. The areas under center pivot irrigation are subdivided using permanent fence to create an inner and an outer circle, and then portable fence is used to strip graze the cattle through the rapidly growing irrigated pastures. This system was designed to make it possible to move the herd quickly from one area to another, while preventing back grazing and overgrazing of the high-quality forage plants.

Figure 14.21. This beef herd is grazing very high-quality forage in this irrigated pasture in central Idaho. Photo courtesy of Jim Gerrish, American Grazing Lands Services.

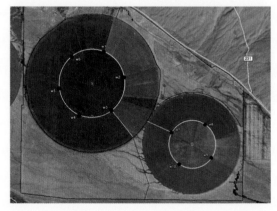

Figure 14.22. This aerial view shows the concentric pattern of circles created by fencing on the irrigated pastures. Photo courtesy of Jim Gerrish, American Grazing Lands Services.

or injuring hooves. Lanes and paddock gates also need to be located so that it is efficient and nonstressful for livestock to exit the pasture through the gate, walk back to the barn, and return.

One innovative solution to managing dairy cow flow between pasture and milking facility is the use of portable parlors. These are unfortunately not widely available or allowed for use in some areas, but they are a very helpful tool that allows dairy animals to be milked and grazed in pastures located a long way from the barn.

WHAT IF ROBOTS MILK THE COWS?

Another technology that has a significant impact on how cows move back and forth to pasture is the robotic milking system. An increasing number of farms are adopting this technology, but it creates some unique challenges for farmers who want to continue to graze the dairy herd. It is possible to continue to graze a herd that is being milked by robots, but the farm must have the right land base, barn location, and grazing system design in order to maintain pasture dry matter intake levels for the herd.

A robotic milking system can offer many labor-saving benefits, but each farm will need to assess whether it is a useful and financially feasible choice. For farms that want to continue to graze the milking herd, or are required to do so by the organic standards, it is critical that the grazing system work well with the new robotic milking system. This section will focus only on the unique grazing-related challenges of cow flow to and from pasture with robotic milking systems.

The robot units are generally located in or right next to the freestall or the loose housing area for the dairy

Figure 14.23. In the Italian and French Alps, dairy herds are grazed during the late summer in high alpine pastures. Many of these pastures are far from the main farmstead, so a portable parlor is parked near the grazing areas. Cow flow is managed with single-strand polywire fences and portable energizers to control where the herd grazes each day, and to bunch the herd close to the parlor during milking.

cows. During the non-grazing season, this allows cows to walk between feed bunks, stalls, or bedded pack areas, and the milking units. Cows can choose when to get milked, when to eat, and when to sleep, and usually have a fairly short walking distance between each area. Farms that have converted to robotic milking systems find that the cows choose to get milked more often than twice a day, which improves milk production and quality. Cows in these systems are calm due to low stress levels.

Each cow has a collar with an ID on it, which the robot reads each time she enters the milking unit. Based on her stage of lactation and level of milk production, the robot gives her some grain and milks her while she eats. Or if the cow has been milked recently, the robot just opens the door and the cow walks out. In addition to getting grain while in the robot, a cow can eat a TMR (total mixed ration) of forages and grains in the freestall or feed bunks. Using the robot to feed the additional grain allows each cow's ration to be individually balanced to her needs. Data on milk production and milk quality of each cow is also tracked constantly. So although installing a robotic milking system is a big

financial investment, there are many advantages for the cows and the farmer.

For those who want to graze the cows, however, these robotic systems often require farmers to significantly redesign their grazing systems. For information on a grazing operation successfully using these robots, and photos, read the farm profile at the end of this chapter.

Cow flow between the pastures and the robots is the primary problem on farms that both graze and use robotic milking systems. When milking takes place in a parlor or tie stall, cows are herded out to pasture as a whole group after each milking. Then later that day, the whole group is brought back in for the next milking. Once a farm has robots, however, cows must have continuous access to the robots. Each robot can milk only one cow at a time, and the robots are designed to have a steady flow of cows passing through. It is not feasible to just bring in a whole herd a couple of times a day for robotic milking. A new grazing system must be designed that allows continuous flow of cows to pasture, to the freestall for additional feed, and to the robot to be milked. For some farms, due to the layout of the pastures and location of the barns, a robotic milking system may not be compatible with grazing the dairy herd.

There are some common principles that all farms with robots will need to consider in designing (or redesigning) their grazing system:

- In addition to the milking robot, robotic exit gates (such as Grazeway's) improve cow flow on most farms. This exit gate reads each cow's ID and then sends her either back to be milked or out to pasture or to the freestall, depending on how recently she was milked.
- Pastures need to be within a reasonable walking distance of the barn so that cows will choose to walk to pasture and back regularly.
- Ideally, a well-designed, well-surfaced cow lane is centrally built to give cows access to the pastures without having to walk over rough ground or through mud. Poor lane surfaces increase the likelihood that cows will either choose not to go out to pasture, or choose to stay out on pasture when they are due to come in and be milked.
- The milking robots are not always able to thoroughly clean very muddy udders prior to milking. (This is another reason that lanes must be well designed to keep the cows clean as they walk back and forth to pasture.)
- Pasture quality must be excellent, and the cows need to know they are getting fresh pasture frequently. Some farms use a system so that they graze three pastures each day. This allows cows to be directed by the computerized exit gate to a new paddock every time they leave the robot/freestall area to go to pasture. These cows know that each time they walk to pasture they will be going into fresh new delicious forages.
- Unless there is a road underpass allowing continuous access to pasture and the robots, road crossings are a problem for most farms once they put in robots. Without an underpass, the only way to graze the other side of the road is to herd a group of cows across the road to the pasture and lock them in the pasture for a period of time, during which they won't be milked. Some farms have found that they can do this with just the "low group" of lactating cows.

- Supplemental feed in the barn must be balanced carefully with the pasture. If the pasture is lush and high in protein and high-protein forages or grains are fed in the barn, the cows will be less likely to go out to pasture to graze. If the grain fed in the robots is changed to a higher energy content, cows will crave the higher-protein pasture and be more likely to make the walk out to it.
- Most farms find that they need to sort and move cows at least once a day to make sure the cow flow is going well. The computer is able to tell the farmer which cows have gone out to pasture through the exit gate, which ones have not been milked recently, and a lot of additional useful management data. With this information, the farmer can go out to pasture and find the few cows that need to come in and be milked, or find the cows in the freestall that need to get kicked out to pasture.

Over time, farmers who are grazing and using robots find that the cow flow improves so that cows are accessing both high-quality pasture and the robots at the right intervals.

Key Points of Grazing System Design

When you're setting up a new grazing system, it can be overwhelming to design, pay for, and build the fence and water system all at once. Starting with larger paddocks that can later be subdivided, letting the animals walk a bit farther for water, or even hauling water to the animals are practical strategies to employ in the first few years.

Over time, lanes can be built to keep animals from walking through pastures, water can be added to individual paddocks, and pastures can be subdivided into smaller paddocks. Create a 3-year or even a 10-year plan to build the fences, lanes, and water system, and to improve pasture quality. This longer time line is a good way to avoid being too stressed out to enjoy the process of using livestock and good management to improve the land. As you create your time line, keep these key points in mind:

- Choose high-quality cropland rather than marginal land when adding land to your grazing system, especially for finishing lamb or beef or grazing dairy cows.

- Select and properly install a quality energizer. An energizer that is not adequately grounded is more likely to be damaged by lightning, will provide less of a charge to keep the livestock in, and is more likely to result in stray voltage.
- Consider subdividing permanently fenced paddocks or larger permanently fenced areas into smaller paddocks with temporary fences, or strip graze the herd by setting up movable front and back fences.
- Situate lanes on high, dry ground. Improvement and maintenance may be needed in muddy wet areas.
- Set up paddocks to separate fast-growing areas from slow-growing areas whenever possible.
- Consider topography and contours — design your layout for south-facing slopes in one paddock and north slopes in another.
- Include shade in some pastures if your local climate includes spells of hot, dry weather. In cool weather, however, it is advantageous to minimize shady loafing areas, where animals tend to congregate and concentrate manure.
- Try to provide water in each paddock.
- Check with your certifier (if your farm is certified organic) about required buffers between pastures and any adjacent non-organically managed land. See appendix D for more information on certification.
- Install gates at the corner of the pasture closest to the barn.
- Consider the impact of an irrigation system as you design paddock shapes and sizes.
- Take advantage of modern cost-effective electric fence technology, lane-building techniques, and portable water troughs.
- Talk to the local conservation district or NRCS office to see what programs there are to assist with design, management, and cost of improvements.
- Attend pasture walks, read some grazing publications, and visit other grass-based farms to learn from their experiences.

The Art of Good Grazing

Grazing with a Robotic Milking System

Hall and Breen Farm in Addison County, Vermont, already had a well-designed grazing system for their organic herd of Holsteins when they installed milking robots in 2011. They were the first grazing farm in Vermont to put in robots, so they had to figure out how to adjust their grazing system to work with the changes to cow flow that robots create.

The farm has two robots that milk a herd of over 100 cows. The soils are somewhat challenging heavy clay, and they have almost 160 acres of land for the milking herd to graze. They take a first cut of hay from some of that land and add the additional land into the grazing rotation once it has regrown to the correct pregrazing height. More land for grazing is available on the other side of the road from the barn and other locations. However, it's too difficult to provide the continuous walking-distance access to those parcels for the lactating herd now that they have the robots, so they make stored forages from those fields.

Farmers Jen Breen and her dad, Louis Hall, have invested in many pasture improvements to make the cow flow work well and improve pasture quality. High-tensile perimeter fence and well-built animal walkways

Figure 14.24. Cows walk into the robot from the other side. The milking unit can be seen on the lower right of this photo, and will move itself to first clean the udder, then milk the cow while she eats some grain.

make moving cows easier. They have also invested in soil amendments and pasture seeds. The farm is certified organic, so the fertility inputs can only be natural materials that are approved by their certifier. They have used wood ash, rock phosphate, chicken manure, and dairy manure. Frost seeding has also been done regularly to maintain plant density and legume content. Over the years the time and money invested in pasture improvements have resulted in some very high-quality pastures with high diversity and density.

The transition from milking in a parlor to using robots to milk the cows had some rough spots as Jen and Louis figured out how to make the cow flow work. The robots

were installed during the non-grazing season. During that first winter, the cow flow with the new system went very well. However, once the first grazing season started, a variety of problems arose with cow flow to and from pasture. Late-lactation cows sometimes chose to stay out on pasture, so milking frequency dropped and SCC (somatic cell count) went up. Jen and Louis had to go out to pasture to bring in cows more frequently while they worked on solving the cow-flow problems.

The changes they made included building new, improved lanes. This solved some problems with mud and rough lane surfaces, which had discouraged the cows from walking on them. The main access lanes also provide some

Figure 14.25. The exit gate sorts the cows as they leave the barn area, only allowing those that have been milked already to walk out to pasture.

shade as the cows choose to walk out to pasture or back to the barn on their own schedule. They also installed a computerized Grazeway exit gate, which doesn't allow cows to go out to pasture unless they have already been milked. They still need to go out to pasture to bring a few cows in, but the cow flow is now working well on the farm.

Jen and Louis say they would definitely not go back to milking in a parlor. The cows would probably agree, too, since they are producing plenty of high-quality milk in a low-stress handling system. They are calm and friendly as they walk between the pastures, the barn, and the robots.

CHAPTER 15

Putting It All Together

Once a grazing system has been designed, there are still many choices to make about how to manage it. Decisions include how long to leave the herd in each paddock and how long to let each paddock regrow. Is it best to use high stock density and let the herd trample the leftovers? Or would it be better to use a lower stock density and clip when needed? Will you supplement with grain and hay or choose to feed your animals only pasture? Is the best plan to strip graze by moving polywire several times a day, or to leave the herd in each paddock for a longer period of occupation? Should you let the lambs and calves walk under the fence to creep feed ahead of the herd or keep them with the mothers by adding a second strand of wire?

The Goal, Plan, and Management Tools

Each farm requires its own unique grazing management system. Systems that are not well designed will appear to work well for a few years, but over time will result in poor pasture, livestock, and soil health. Other systems will create problems that are visible more quickly.

The ideal grazing system is planned to meet the farmer's quality-of-life goals as well as goals for the livestock and pastures. This well-managed system will be what creates ideal livestock performance and improves pastures over time. Grazing system infrastructure alone cannot create these improvements; a well-thought-out plan based on knowledge of plant and animal needs is essential to avoid common grazing mistakes.

Having a clear goal for the farm business can be very helpful in deciding what intensity of management is ideal. This should include consideration of how much human labor is available to move livestock and fence, what the animal performance goals are, and what the goals are for

improving the soils, plants, and infrastructure. The goal should also include a description of the quality of life that you, the farmer, would like to have and what you value most. This can be very helpful in assuring that the grazing system is creating the environment you want to live in, while also designing it so that it's not creating more work than you can do.

If the goal is to improve pasture as quickly as possible, then using higher stocking densities and more frequent moves will be best. If the goal is to gradually improve pasture quality while still having time for an off-farm job, then a system where the herd is moved once every few days or even once a week will work better. If the goal is to provide all or most of the forage for a high-producing group of cattle, then in addition to rapidly improving pasture quality it will be necessary to stay focused on maximizing dry matter intake of high-quality pasture.

There are so many ways to creatively design and manage a grazing system that it is not possible to cover all of them in this book. In this chapter, I discuss how you can use "tools" such as stock density, timing of when to graze, grazing multiple groups, trampling, and clipping. The Art of Good Grazing sidebars, which you can find at the end of most chapters in the book, are real farm examples of creative and effective grazing systems. The conclusion of this chapter and appendix A also include important information on how to monitor pasture ecosystem health to determine whether your grazing system design and management are improving the pasture or not.

INCREASING STOCK DENSITY

To increase stock density, paddock size has to be smaller so you are grazing more pounds of livestock per acre. However, the animals' nutritional needs must still be met, so unless there is some way to significantly increase

Figure 15.1. This farm is using a higher stocking density to trample seed stems of cool-season grasses in early summer. Because the herd is moved frequently, the livestock are able to get enough good-quality forage while also having a useful mob/trampling impact.

the amount of dry matter per acre, you will have to move the animals more often.

As discussed in previous chapters, frequent herd moves are commonly accomplished by using portable fence rather than building more fixed, permanent paddocks. Portable fence for interior subdivisions, in combination with permanent perimeter fences, provides the most flexibility for paddock sizes, stock density, and frequency of moves. This type of infrastructure design also allows flexibility for varying herd sizes due to growing lambs or calves or seasonal changes in livestock numbers.

This system also allows flexibility to plan a vacation into the annual grazing plan. During vacation, larger paddocks can be set up, and a neighbor can move the herd to a new paddock every few days or even once a week.

Some farms prefer to use permanent fence only on the field perimeters and use portable fence for all the interior subdivisions. This allows maximum flexibility but requires more labor during the grazing season. Others prefer a system of paddocks sized to each provide a planned number of days of feed for the herd. These fixed-size paddocks can be strip grazed with polywire so stock density is high, or the herd can be left in each one for a few days. The downside of using fixed paddocks that are each grazed for a few days or a week is that clipping may be needed to control weeds or remove excessive rejected forage. The benefit is a less daily labor-intensive system, which, as noted on page 189, may be ideal for farmers who have off-farm jobs.

VARYING PADDOCK GRAZING ORDER

A common grazing mistake is to graze a set of paddocks in the same order every year, and then continue to regraze in that same order throughout the grazing season. Because it is very likely that each paddock grows at a slightly different rate at different times of the year, some will need longer or shorter rest periods. If they are always grazed in the same order, some paddocks will suffer overgrazing damage.

A better approach is to take a tour of the pastures at least once a week to see which ones are growing rapidly or slowly. Also note which ones have problem weeds that need to be grazed, mowed, or trampled before they produce seed and spread. Use that information to decide what order to graze the paddocks in. Once you have information on how different parts of the farm grow, you can use it to create your annual grazing plan, replanning during the grazing season.

In addition, if the same paddock is grazed first each spring, all the plants in that paddock that grow most vigorously in cold spring soils will be at a disadvantage. Every spring the fastest-growing plants in that spring paddock will be grazed off early, while the slower-growing plants aren't tall enough to be grazed yet. Over time this paddock may lose its fast-growing taller plants and become dominated by shorter, slower-growing plants. So while it's easiest to always start grazing the paddock closest to the barn, the plants would prefer it if the animals started at the far end of the farm every other year so they could get a little extra rest!

This is another example of why it's important to have a grazing plan and update it yearly. Plan ahead to start the grazing rotation on a different part of the farm each spring so that the livestock don't selectively graze the vigorous spring plants each year in the same place.

This same idea should be applied to fall grazing. While it's tempting to have a set routine of where to graze the herd each fall, it is better for the plants to plan ahead and stockpile different parts of the farm. This allows the plants to have a chance every other year to go into fall with more residual and root reserve.

LEADER–FOLLOWER GRAZING

Some farms use a follower group of cattle that have lower nutritional needs to "clip" the pasture left behind by a leader group. In this system, a group such as dry cows and older heifers will graze behind the milking group. The milking cows, which have the higher nutritional needs, get the most digestible and palatable forage in the pastures. The follower group then grazes and tramples what is left. This can reduce the need for clipping, while letting the milking group graze through an area quickly, eating only the best forage. However, when using this method, care must be taken not to let the followers graze the plants too short or cause overgrazing damage. If the animals graze the plants too short, not only will they damage the plants, they also run a higher risk of being infected with internal parasites.

The follower group needs to be moved into the paddock quickly once the leader group is moved out so that the plants don't have time to start to regrow. Additionally, the follower group can't be left in the paddock for too long, or they may graze the plants down too short or graze them again just as they start to regrow. When done correctly, leader–follower grazing systems can have many benefits for pastures and livestock.

Leader–follower grazing isn't a great match for every farm. It requires well-fence-trained animals and a hot fence, so the two groups don't jump the fence to hang out together. Leader–follower systems generally work best when water is available in each paddock so that both groups can be locked in their respective paddocks with no need for access to a shared lane or water tub.

Figure 15.2. At Does' Leap Farm in Vermont, sometimes the horses are used as a follower group behind the goats. The farmers take care not to let the horses graze the plants too short or for too long after the goats finish grazing the area.

CLIPPING

Mechanical clipping of pastures with a brush hog or mower can be a great strategy in certain situations. However, it should only be done when necessary since it takes time, fuel, and wears out farm equipment.

Farms that are able to consistently graze at just the right pregrazing height with high stock density or with a follower group may never need to consider clipping. Farms with animals that have lower nutritional needs may also find that clipping to improve the quality of regrowth isn't needed. Many classes of livestock can do just fine on pastures that include some grass seed stems and a few weeds.

Clipping can be useful and worth the time for dairy farms where maximizing intake of highly digestible pasture throughout the grazing season is important, and on farms where there are problem weed species. For farms grazing cool-season perennial grasses with high-performance animals, a well-timed post-grazing clipping to remove seed heads can be worthwhile.

Because cool-season grasses make their seed stems and seeds in late spring or early summer, once clipped off they will produce mostly leaves for the rest of the grazing season. The best time to clip is immediately after grazing to make sure no new growth is being mowed off. If you wait too long after grazing to clip, and any new regrowth is mowed, it's as if you're doing overgrazing damage to the plant! So ideally the paddock should be mowed right as the animals move out. Then the next time the herd is moved into this paddock, it will present a nice uniform pasture that has more leaves and fewer stems.

Another reason to clip is to remove weed species. Clipping can be particularly helpful in preventing the spread of weeds, if it's possible to clip the weeds just before they make seeds.

Remember that this post-grazing clipping is only needed in areas where there are a lot of leftover standing weeds, stems, and plants. Paddocks that were grazed with a higher stock density, with less selective grazing and more trampling, won't need it. So only clip if you have to and consider using a higher stocking density or leader–follower group to save on fuel and equipment usage instead.

So far we have only discussed post-grazing clipping. Preclipping is another grazing strategy that, in some situations, may increase pasture dry matter intake. The theory of preclipping is that livestock will readily consume more when plants, which would otherwise be less palatable, have been mowed and allowed to wilt. Farms using this strategy will often clip some or all of a paddock where the plants have grown too tall or otherwise become less palatable. They then let the mowed pasture wilt for part of a day and turn the herd into the area to graze it.

While some farms found that feeding the wilted preclipped pasture was helpful, others found it did not work well for them. These farmers found that their dairy herds did consume the preclipped, fibrous seed stems, but that this less digestible forage resulted in poorer cow performance. As a result, these farms have shifted back to post-clipping or following the dairy herd with a group of animals with lower nutritional needs (dry cows or heifers).

When clipping before or after grazing, keep in mind that the height of the mower is important. If the mower height is too low, the crowns of the pasture plants can be damaged. Pastures that have been repeatedly clipped too short will lose plants such as red clover, alfalfa, timothy, and orchard grass and gradually shift to a shorter-growing sod such as Kentucky bluegrass and white clover.

WHEN TO START AND END THE SEASON

Spring pasture workshops often include lively debates on when the ideal time is to start grazing each spring. Some prefer to start as early as possible to try to keep pastures from growing too rapidly later and becoming overmature. Others prefer to wait until the pasture height is tall enough that the cattle can easily get enough dry matter intake and all the pasture plants have had time to fully regrow after winter dormancy.

There are pros and cons to each approach. The best thing to do is to make sure your choice is well thought out and written into your grazing plan. If you decide to start grazing early to try to maintain a supply of highly digestible pasture, keep in mind that the plants will do better over time if you begin the early-spring grazing on a different part of the farm each year. If you prefer to allow more growth before grazing, it's important to consider some additional early harvesting or clipping of excess pasture in order to maintain forage quality once grasses begin to form seed heads.

In addition to keeping the needs of the plants in mind while planning spring grazing, consider the needs of the livestock. The transition from stored forages being fed in a sacrifice pasture, barn, or heavy-use area to providing all or most of the dry matter from pasture requires a behavioral shift as well as a significant change in what is happening in the ruminant digestive system.

Allowing a gradual shift from stored forage to pasture over a week or two is ideal. This can be done by limiting the number of hours the animals have access to pasture each day, or by continuing to provide some supplemental forage on the pasture.

Fall grazing, if not well thought out, can damage plants in several ways. Allowing animals to graze plants that are not fully regrown in the fall depletes energy reserves that are needed for winter. Additional damage is done in fall if plants are grazed for too long as they are regrowing, or animals are allowed to graze them down too short. Any of these fall grazing mistakes can significantly reduce next year's pasture productivity. Careful planning of where to graze in the fall as plants move into winter dormancy is important to prevent this from occurring.

Fall grazing should be planned so that plants still have plenty of time to regrow before each grazing, even though growth rates may become very slow as temperatures drop or if soil moisture is lacking. It is also important to let those plants go into the winter dormant months with plenty of energy reserves. As discussed in part 2 of this book, overgrazing is particularly damaging in the fall when many plants need to store energy reserves in their roots and the bases of their stems. They need this energy to survive the winter and grow again in the spring.

It may be tempting to put livestock into pastures in the fall to "clean up" by grazing everything. The results may look tidy, but this practice does a lot of damage to the pastures. Remember that good-quality pasture is going to look messy but will be more productive.

MULTISPECIES GRAZING

Grazing sheep or goats as well as cows can provide advantages but also many challenges. The benefits of a well-planned multispecies grazing system can include reduced problems with internal parasites, improved forage utilization, and the ability to use different animal groups to create changes in pasture species. Challenges of working with both small and large ruminants include the needs for different types of fence and the additional labor of managing more animal groups.

If the goal is to use cattle grazing to reduce parasite problems in sheep or goats, note that just putting the animals together in the pasture in one group, or running them as a leader–follower grazing system, will *not* solve parasite problems. Using multispecies to control internal parasites requires a grazing plan that includes an understanding of the parasite life cycle (see chapter 12 for more detail).

An effective multispecies grazing system to manage parasites requires knowledge of which pastures have the highest concentrations of infective parasites, and how long the parasites can continue to survive in the pasture. A forward grazing system that keeps the most susceptible animals rotating into clean, non-infected pasture can help manage parasites. By harvesting crops of hay, growing annual crops, or using dairy or beef cattle to graze pastures after the more susceptible goats and sheep have grazed, it is possible to reduce parasite infectivity rates in pastures.

Multispecies grazing can also be used to control weed species. Careful use of the different grazing and browsing abilities of goats, sheep, and cows can be done to apply more grazing pressure to certain plant species. If you stress some plants, such as weeds, via trampling, severe grazing, or more frequent grazing, you can reduce or even eliminate them from the pasture.

Goats, for example, can be used to browse even thorny vegetation such as multiflora rose multiple times in a season. The rose plants can be killed within one or two seasons if many of their leaves are removed multiple times per year. Another example would be the use of a beef herd to trample an area of problem plants. By using a high stock density of these heavier herbivores, problem plants can be damaged and more desirable grass and legume seeds can be planted and have an opportunity to germinate and thrive.

Grazing Record Keeping

Records are another important tool for graziers to use. Keeping records of how much feed there is in a pasture can help you more clearly see trends in forage production and changes during the grazing season, which assist in long-term planning. This information can also highlight problems with soil fertility, soil moisture, or plant species selection. It's much easier to find a problem with pasture dry matter production and fix it if it's being measured!

If you take a weekly measurement of pasture dry matter per acre from each paddock, you can graph out this information to create a visual of the amount of forage on the farm. This is called a grazing wedge. The University of Missouri has a computer spreadsheet that can be used to make a grazing wedge record and monitor pasture dry matter. Some farmers use this system to plan where to graze, estimate how much more forage is left before the end of the grazing season, and even decide where to spread fertilizer.

For a farm that is certified organic, or in an NRCS grazing program, some types of records are required. In addition to meeting these requirements, grazing records can also provide helpful management information on how much forage the pastures are providing and whether the pastures are improving or getting worse. Previous grazing records are also helpful as reference when planning out the next grazing season.

The ideal pasture-planning or record-keeping system is one that you will actually use. Since each farmer has a different way of thinking and different preferences for record keeping, the record system on each farm will need to be a good match for your needs and personality. A fancy record-keeping system that no one uses won't be helpful!

Records may be as simple as notes on a calendar or in a notebook about how many animals were in the grazing group or groups and which fields they were in on different dates. A laminated copy of the farm map can be kept on the farm office wall and dry-erase marker used to note the dates each pasture was grazed. At the end of each grazing season this information can be recorded by simply taking a digital photo of the map on the wall, and then the map can be reused the next grazing season.

Holistic Management International or your grazing consultant can provide grazing planning charts and teach you how to use them as part of a planned grazing system. These charts both lay out the plan and record where the cows actually graze and how long they stay in each area.

This proactive planning process is a great way to make sure that pastures are not grazed too often, that hay crops or other ways to deal with excess forage are planned, and that pasture stockpiling to extend the grazing season is done correctly. These planning charts also allow farmers to plan a summer vacation, keep track of wildlife activity such as bird nesting, and maintain records of things such as rainfall.

Monitor and Replan as Needed

It is important to have a grazing plan, but also to constantly monitor the *actual* plant regrowth rates in pastures, as well as the weather and the livestock. For example, every spring is different, so even though the plan was to start grazing on April 15, if it's a cold year it may be better to replan and start on May 1. In addition, the plan may require some revision due to too much rain, not enough rain, plant disease, pest problems, or an unexpected family event that leaves labor in short supply.

Using a grazing chart that allows the original grazing plan and the actual grazing activities to be recorded on the same page is ideal for this grazing planning and replanning process. Over several years this information can be saved and used to create more accurate and flexible grazing systems that better meet the needs of the farmers, the livestock, and the pastures.

A pasture may look nice from the road or lane as you drive by. However, if you get out of the truck and walk through the pasture, you may see problems you could never spot from the road. So the monitoring process needs to include time spent regularly walking pastures to make observations of the soils, plants, density, and growth rates.

Some observations show short-term adjustments that need to be made. These include things such as changes in herd size or their dry matter needs. But some longer-term monitoring of the pastures is also important so that the results of the grazing management system can be seen. Some of the key factors that can be helpful to monitor include plant cover, plant residual, diversity, plant vigor, litter, soil fertility, soil health, and erosion. Appendix C in this book has a simple pasture monitoring worksheet to use. More comprehensive pasture monitoring training and worksheets are available from NRCS and Holistic Management International. These methods, particularly when done annually, are a great way to track pasture changes over time.

MONITORING THE PASTURE ECOSYSTEM

Essential processes in the pasture ecosystem include the water cycle, mineral cycle, energy flow, and community dynamics.[1] It's important to pay attention to these processes as a way to make sure the grazing system is improving the health of the ecosystem over time.

The Water Cycle

Water is constantly being cycled among atmosphere, ocean, and land. As water evaporates, vapors rise and condense into clouds. The clouds move over the land, and rain, ice, or snow falls on the land. Some water runs off into streams, rivers, lakes, or oceans. Some penetrates the soil surface and becomes bound to soil particles or flows down to replenish groundwater supplies. If the soil is bare and uncovered by residue or plants, even tiny raindrops may cause some damage to the soil structure. Soil that is uncovered or has poor structure is also more likely to erode.

For pastures, the more water infiltrates into the soil and is retained, the more is available to the plants. By increasing soil organic matter and improving soil structure, more water can be held in the soil. The water that is retained in the soil will be drawn up plant roots and eventually transpired into the atmosphere so the cycle can

continue. However, it is also important for the soils not to become too saturated; plants can "drown" if soil conditions are too wet for so long that they become anaerobic.

The function of water in soil is critical to plant and animal life. Creating an effective water cycle will encourage productivity both above- and belowground. An effective water cycle allows water to infiltrate into the soil and become available for plants and soil organisms. An ineffective water cycle allows water loss from the soil surface in the form of runoff or evaporation, and plants have minimal opportunity to utilize the water. An effective water cycle helps buffer the erratic nature of rainfall, making floods and droughts less severe. An ineffective water cycle increases the frequency and severity of drought and flood events. Monitoring the effectiveness of your water cycle is easy. While walking your land, look for the key identifiers listed in table 15.1.

An Effective Mineral Cycle

Like water, minerals flow in a cyclical pattern. Living organisms continually release nutrients from their bodies in the form of wastes, which are by-products or leftovers from metabolism. One organism's wastes become another organism's nutrients. This process is called mineral cycling.

Every living plant and animal on earth is made of nutrients that have been used and reused over and over again. In the context of a grazing system, this means that an effective mineral cycle will continually cycle and provide the essential nutrients needed by plants,

Table 15.1. Pasture Water Cycle

Signs of an Effective Water Cycle	Signs of an Ineffective Water Cycle
No bare soil exposed	Excessive soil exposure
Soil is crumbly	Soil not crumbly and is instead capped or sealed
Soil around plants is covered with fallen plant debris	Soil bare around plants
Diversity of plants and root structure	Lack of plant diversity and root structure
Plants start to grow earlier in the growing season, and continue growing longer even into fairly long dry periods	Plants start growing later in the growing season and stop growing earlier in the season; dry periods are extremely unproductive
No signs of surface water flow	Signs of surface water flow (sheet erosion, rills, silt deposits, grass roots sticking above soil surface)
Water levels in springs or wells are stable	Water levels in springs or wells are dropping

soil microorganisms, and animals. For a mineral cycle to operate optimally, the soil must be biologically active (microorganisms and roots) with ample aeration and water (effective water cycle).

Plants are responsible for lifting nutrients from the soil to the surface and making them available to animals. Plant roots and soil organisms extract minerals from the soil. A correctly functioning mineral cycle will include plants that have healthy, large root systems as well as root systems with varying structures. Plant and animal materials containing nutrients from the soil are eventually returned to the surface in the form of dead plant and animal residue (organic matter). At this point, the minerals in this plant residue are not available for reuse by plants. This material is broken into smaller pieces by surface-feeding insects and other soil organisms. The minerals will be incorporated into microbial biomass, and their by-products (various nutrients) will become available to plants. Nutrients contained in the microbial population will be released for plant uptake as they die and are decomposed by other soil organisms. To keep the soil biologically active, the living organisms must also have ample moisture and air. This is possible only by maintaining the condition of the soil surface so that soil can retain moisture and air.[2] Some key identifiers to an effective mineral cycle are listed in table 15.2.

Effective Energy Flow

In general, energy flow begins with sunlight landing on leaves, where it is converted by photosynthesis into forage. The forage is then converted by grazing or harvest into meat, milk, or fiber by the livestock. This was discussed in more detail in Maximizing Sunlight Conversion in chapter 3.

How well energy flows through a farm depends on how sunlight fuels the farm's biological systems. Sunlight lands on a green plant, allowing it to grow. The plant is eaten by an animal, which is then eaten by a predator, and so on. Eventually the plants and animals die and are consumed by decomposing organisms. In this process, energy is taken underground by roots as well as by the decay process.

Table 15.2. Pasture Mineral Cycle

Signs of an Effective Mineral Cycle	Signs of an Ineffective Mineral Cycle
Ample litter on surface of soil	Little to no litter on surface of soil
Many surface insects are visible	Few to no insects are observed on soil surface
Litter and animal manure decomposes quickly	Litter and animal manure decomposes slowly
Soil surface is porous and soil is crumbly with good structure	Soil surface is crusted or sealed and soil lacks crumbly structure
Diversity of plants and root structure	Lack of plant diversity and root structure
Plants have healthy, large root systems	Plants have diseased or small root systems
Plants appear healthy and do not show signs of nutrient stress	Plants not vigorous and show signs of nutrient deficiencies
Soil smells earthy	Soil has no smell or does not smell earthy

Table 15.3. Pasture Energy Flow

Signs of Effective Energy Flow	Signs of Ineffective Energy Flow
Close plant spacing, high plant density, and high leaf area index	Bare soil visible between plants
Rapid plant growth	Slow plant growth
Plants growing over a long growing period	Short season of productive growth
A long food chain, including grazing animals, predators, and decomposing organisms	A short food chain with fewer grazing animals and low biological activity of decomposing organisms

An effective energy flow on a farm depends on having plenty of total area of actively growing leaves that can convert sunlight into more plant growth as efficiently as possible, and for as long a growing season as possible. Energy flow is also dependent on how forages are grazed, harvested, and decomposed. Monitoring energy flow on the farm involves finding where solar energy that *could* be captured to create more life is not being used. Some key identifiers to an effective energy flow are listed in table 15.3.

Effective Community Dynamics

Community dynamics are improved on a farm as the level of diversity rises.[3] As the complexity of the community of plants and animals increases, the land's stability, health, productivity, and resilience will also increase. Understanding this can be aided by thinking of the plant community in terms of succession — the stages or steps of plant community development. Generally, bare ground is first replaced by moss, algae, or lichen, then this converts to annual forb or grass species, then more perennial grasses, then to brushy species, and then to forest.

This is a gradual process that involves an increasing diversity of species and increasing amounts of biomass. Farmers encourage this process by planting diverse mixtures of plants instead of monocultures, practices to replace annuals with perennials, and by using good

Table 15.4. Pasture Community Dynamics

Signs of Good Community Dynamics	Signs of Poor Community Dynamics
Many different plant species	Fewer plant species (monoculture)
Many perennial plant species	Increasing numbers of annual plant species
Decreasing amounts of bare soil visible	Increasing amounts of bare soil visible
Rapid manure decomposition	Slow manure decomposition

grazing practices to increase plant diversity. When monitoring community dynamics, look for the factors listed in table 15.4.

PASTURE MONITORING RECORDS

Keeping a written record of observations can be helpful. This record of pasture monitoring will allow you to keep track of the changes that occur due to management over time. Records can be as simple as some digital photos taken each year in the same place on the farm. They can also include more detailed pasture condition scoring checklists from NRCS[4] or comprehensive biological monitoring tools and training from HMI.[5] Refer to appendix C for a simple pasture monitoring record sheet.

The Art of Good Grazing

Grazing a Large Dairy Herd

At the Horizon Organic dairy farm outside Kennedyville, Maryland, 480 lactating cows are grazed in a carefully designed and managed pasture system. Providing this many cows with high-quality pasture requires skills at both herd and pasture management.

The farm managers and staff have been well trained in managing cows and pastures, and the farm has carefully invested in cow lanes, miles of electric fencing, a pasture drinking water system, improved pasture plant species, and a pasture irrigation system. The pastures are on high-quality soils and have been managed to continue to build soil health and fertility.

The farm's successful grazing system is based on four key factors:

- Grazing and paddock management.
- Herd and farm investment.
- Well-designed infrastructure.
- Efficient farm management.

GRAZING AND PADDOCK MANAGEMENT

The goal of the Horizon farm's grazing system is to provide high-quality pasture to the herd while also using the cows to improve the pastures. This requires moving the grazing groups frequently into new pasture. Each new area is then grazed for a short period of occupation, often two days or less. This lets cows eat about one-third of the pasture and leave the rest behind, meaning the herd can eat the highest-quality forages without damaging the plants by grazing them too short or too often. This is a win–win management system for both plants and cows.

During the grazing season, cows are fed a portion of their daily ration in the freestall as a mix of forages and grain. They get the rest of their daily dry matter intake requirements from pasture. The lactating herd is milked and grazed in four separate groups of about 120 cows, allowing the staff to manage cows with different needs separately. It also ensures that each smaller group doesn't spend too much time walking back and forth to pastures or waiting to be milked.

Figure 15.3. This lane at the Horizon Organic dairy farm divides perennial grass legume pasture on the right from annual sorghum sudan grass pasture on the left. The lane is constructed to hold up to heavy hoof traffic. The fence carries voltage efficiently over long distances to provide sufficient charge for polywire subdivision fences throughout the pastures.

During the non-grazing season, cows are milked three times a day. But during the grazing season, part of the herd is milked just twice a day, allowing them to spend more time on-pasture and reduce the distances they need to walk. During the grazing season, the two groups that continue to be milked three times a day graze paddocks closer to the barn. The other two groups milked twice a day walk farther to graze.

Some groups must walk more than a mile and a half round-trip to reach some pastures. During hot, humid summer weather, cows graze only at night, and stay inside during the day with fans and sprinklers to prevent heat stress and maintain feed intake.

Pastures include annuals such as sorghum sudan grass and triticale as well as perennial pastures with a high legume content. Legumes provide nitrogen to other plants in the pasture, a particularly important nutrient for an organic farm since synthetic nitrogen fertilizer is prohibited. The high-legume pastures also provide a significant amount of the dairy cows' protein needs during the grazing season.

In previous years tall fescue was planted in a lot of the pastures. However, due to the availability of irrigation or

Figure 15.4. Cows grazing a perennial grass legume pasture in the springtime look back over the fence at what they grazed the day before. The cow manure in the foreground is typical for animals grazing these lush, high-legume pastures.

Figure 15.5. After only seven days, this pasture is regrowing rapidly. It will be allowed to grow for at least another two weeks before the cows graze here again.

rainfall, they have found that they are able to grow much more palatable cool-season grasses and legumes that the cows prefer to graze — so they have gradually been reseeding the tall fescue paddocks to more palatable perennial mixtures. By rotating through an annual crop such as sorghum sudan grass for a season, they can kill the fescue sod so that the new perennial seeding is able to get established successfully.

In the past, the farm set up smaller paddocks in order to increase stock density and thereby increase the herd trampling effect. However, the cows didn't do as well under this system, so they have increased the paddock sizes. Cows are still given fresh new plants each time they go out to pasture, but the paddocks are larger and any residual that the herd hasn't trampled is clipped if needed. With the slightly larger paddocks, and lower stock density,

they are able to manage the pastures well while providing quality feed and low-stress herd handling.

HERD AND FARM INVESTMENTS

After each grazing/clipping cycle at Horizon Organic dairy farm, the paddock is left to regrow for three to five weeks. The length of the regrowth period depends on how fast the pasture plants are growing and is managed carefully so that animals don't return to the paddock too soon.

Moving four groups of cows to a new pasture frequently, and then letting each area regrow for more than three weeks, can quickly add up to a large number of paddocks and fence in use on the farm. This system also requires drinking water in each paddock and lanes that can handle a lot of hoof traffic. Multiple groups of cows use the same lane, so once each group is grazing

Figure 15.6. Cows are about to be turned into this paddock, which has been planted with sorghum sudan grass. Because of the large herd size, two water tubs have been set up to make sure there is always enough water.

the gate is closed so they stay in the paddock while other groups can use the lane. In addition, during dry periods it is necessary to run irrigation, and when there is excess forage or weeds mowing is required.

Fencing is done with high-tensile permanent fence along the lanes and around the perimeter of large fields. Portable polywire and posts are used to create interior paddock subdivisions. The use of the portable fence allows flexibility in paddock sizing as the grazing group sizes change throughout the grazing season.

WELL-DESIGNED INFRASTRUCTURE

Drinking water is provided by two large, portable tubs with float valves in each paddock. Water lines are buried under

the lanes so that these portable tubs can be connected to a high flow rate of water anywhere on the farm. Lanes required good design and solid construction to withstand four groups of cows walking back and forth to the barn multiple times each day. To limit these impacts, most lanes have road-building fabric under a solid layer of fine gravel.

The majority of the pasture on the farm can be irrigated when necessary using two center pivots. Without irrigation, the farm would have to allow much longer regrowth periods for pasture plants during dry weather. This would make it much more challenging to provide enough high-quality pasture to so many cows. However, the logistics of irrigating a pasture that has been subdivided into many small paddocks are not always easy. Taking down the fence

Figure 15.7. Because the size and shape of the pastures on the farm vary, different lengths of polywire are needed in different areas. By having a labeled farm map in the office, and labeling each polywire reel, farm staff can make sure the reels are in the right place the next time the paddocks are set up.

each time the irrigation is run isn't practical. To solve this, several modifications to the fences were made that allow the wheels of the center pivots to run over the fences instead of having to move fence out of the way.

EFFICIENT FARM MANAGEMENT

To keep track of feeding, milking, and herd movement for each of the animal groups, the farm uses an extensive record-keeping system. Farm maps, which also include names for all the paddocks, are posted in several locations on walls in farm offices. This makes it easier for farm staff to communicate with one another about which paddock each group grazes in, as well as which areas are being clipped, irrigated, or allowed to regrow.

The farm keeps extensive records of just about everything, including details about which pasture is grazed by which group, along with milk production and feeding records. By closely monitoring pasture quality and cow performance, the farm managers are able to adjust the supplemental ration fed to each group every day. For example, when cows are getting more protein from a high-legume paddock, protein in the ration is reduced and more energy is added.

These records are kept primarily as a tool for the farm staff to manage cow performance and well-being. However, they also play an important role in maintaining the farm's organic certification, which includes regular inspection by an organic certifier.

Both the cows and pastures are doing well on this farm. Some of the reasons for this success are:

- The staff are following the key principles of good grazing management: short periods of occupation and long, variable recovery periods.
- The pastures are growing on quality cropland.
- The farm has invested in well-designed lanes, electric fencing, and a piped drinking water system.
- Irrigation is available if rainfall doesn't provide enough soil moisture.
- The staff have maintained good soil health through good grazing practices and organic fertility inputs.
- The managers have selected highly palatable locally adapted plant species.
- The staff carefully monitor pasture forage quality and supplemental TMR being fed to each group of cows.

Beyond the Farm Gate

I have been learning about and practicing grass farming for many decades now, and I find it ever more fascinating. My continued interest is largely due to the incredible complexity of the pasture–livestock relationship. It is also because when done well, grass-based livestock farming is a beautiful way to have a positive effect on a parcel of land and on a small part of the planet. Great learning begins with wonder, and there is so much to learn and be amazed by as we understand more about the ecosystem we all live in.

The science of pastures is not just about grass and cows. Understanding pasture involves the biology of rumen microbes and soil organisms, the diversity of plants and their individual niches, the mysteries of soil health, the unpredictability of weather, and the synergistic interplay of all these interrelated parts. When done right, pastures can benefit our planetary ecosystem, and the livestock provide us with nourishing foods. Well-managed pasture also allows us to experience the joy of witnessing livestock on pasture where they are able to more fully express their natural behaviors and thrive.

But within all this complexity, keep in mind that all good grazing systems are based on some underlying guidelines. These basic guidelines have been known for at least a couple of centuries and have been written about by many scientists and farmers. I have repeated them many times throughout the book, but I'll summarize them again here just in case this is the only chapter you read!

In order to keep grazing-adapted plants in the pasture growing well and provide high-quality pasture to livestock, it is essential to use both:

- Short periods of occupation in each paddock.
- Long recovery periods, which are varied based on how fast plants are growing.

Although these principles are very widely known, many grazing systems still fail to do well because the farmers leave animals in the pasture too long, or return them to previously grazed pastures too soon. Knowing — and practicing — the basics is essential!

Of course, there is so much more to learn and observe than just the basics. We continue to learn more about why these basic guidelines of good grazing management can work so well, particularly when combined with new technologies such as electric fencing and new methods of assessing soil health. We are also learning about the benefits of good grazing practices in dealing with our changing climate and other environmental challenges. The increasing body of knowledge shows us the complexity and elegance of the interrelated parts of farm ecosystems and their role in the larger planetary ecosystem. Well-managed grass-based livestock farms are playing a part in improving the environmental health of the ecosystem of which we humans are one component part.

Good grazing practices on one individual farm *can* make a difference. Each grass farm can become a source of healing and regeneration through soil building, carbon sequestration, improved water quality, protection from extreme drought and rainfall events, improved humane animal care, and the production of healthy foods. Perhaps one farm at a time, and one customer at a time, we can turn things in a more positive direction for our shared planetary ecosystem.

Grass farming alone can't solve all the challenges facing humanity. We do need to improve large amounts of poorly managed land and shift from confinement farms to grass-based livestock farms. But we must also address social and economic injustices so that all consumers can afford to buy good food, and so that farmers can be paid a fair price for the products they grow.

Paying farmers enough to allow them to invest in the practices and technology to build soil, improve water quality, and sequester carbon is obviously a necessary part of the solution. In order for farmers to be paid a fair price for their products and their sustainable farming practices, the people buying those products must have enough income to afford them. Economic injustice, lack of access to education and opportunities, and low wage rates force a significant number of people to buy poor-quality cheap food. And the real cost of our cheap food policy is not understood by most consumers.

This cheap food sold at many convenience and grocery stores contains largely commodity ingredients and is highly processed. Eating such food has cumulative negative impacts on human health. This food is not nutritious enough to sustain health and is contributing to an epidemic of health problems.

Somehow we need to make nutritious food available to everyone. But this will require us to raise people up out of poverty, cease our endless wars, and learn how to be caretakers of our ecosystem both in our farming practices and in all our daily activities.

Every person can make a difference, and the synergistic effect of our collective effort will make a positive impact on our unique and lovely planet. Just as the farm ecosystem is a complex mix of interdependent parts, so are our larger local communities and nations. So although we are doing important work by shifting farms to more sustainable management practices, each farm depends on many other parts. Everything is interconnected. Humans have created magnificent art and music. We have learned how to sail the seas and fly into space. Surely we are smart enough to care for the ecosystem upon which we all depend.

Troubleshooting Pasture Problems

P asture problems may include issues with poor livestock performance, poor pasture productivity, low pasture quality, soil health, erosion concerns, and more. Most of these topics are covered in the text of this book. This is a summary of some of the most common issues and what the cause and potential solutions might be.

Table A.1. Livestock Problems

Symptom	Cause/Problem	Some Potential Solutions
Livestock are rejecting pasture.	Pasture is too tall/overmature.	Move cows to pasture with appropriate pregrazing height. If that's not available, strip graze the tall pasture.
	Paddock is too large so they are selectively grazing and wasting too much forage.	Recalculate paddock sizes and make them smaller.
	Supplemental feed in barn contains too much protein so cows choose not to eat high-protein pasture plants.	Reformulate ration to take out or reduce high-protein forages such as haylage and replace high-protein grain with energy grain. Test MUNs.
	Gate to pasture is left open so cows can come back to barn whenever they want.	Shut gate and provide water in each paddock.
Young stock have rough hair coats, poor rate of weight gain, or poor body condition.	Poor forage quality or inadequate quantity.	Increase rest and regrowth period for each pasture before grazing with young stock to assure there is enough pasture grown back to meet both plant and animal needs.
		Test soils and assess plant species, then develop plan to improve pasture quality and productivity.
	Internal parasites.	Get vet to test for parasites and treat if necessary. Create prevention plan, which should include taller pregrazing heights, taller residual, improved nutrition, and grazing "clean" pastures with most susceptible young stock.
Lactating herd or flock drops in milk production and body condition score during summer grazing season. Some or many cows won't breed or stay bred during summer grazing season.	Paddocks too small to provide enough DMI per cow.	Recalculate paddock sizes and make them larger.
	Pasture quality poor.	Sample pasture forages and reformulate supplemental feed in ration, and make plan to improve pasture quality.
	Pregrazing height too short to allow cows to get enough DMI.	Make sure pastures are fully regrown so they are tall enough when livestock go into each paddock.
	Supplemental feed in barn contains too much protein.	Reformulate ration to take out or reduce high-protein forages such as haylage and replace high-protein grain with energy grain. Test MUNs.

Table A.2. Pasture Plant Problems

Symptom	Cause/Problem	Some Potential Solutions
Pasture is becoming more weedy.	Low plant density provides bare soil for seeds to germinate on.	If density is very low, seeding improved pasture species may be needed.
		Correct current and past overgrazing damage:
		• Check grazing management system to make sure plants have adequate time to regrow after each grazing. • Make sure period of occupation is not so long it allows animals to regraze plants that are actively regrowing.
	Selective grazing.	Increase stocking density and decrease period of occupation.
	Invasive weed species.	Prevent spread by grazing, mowing, or trampling before seeds form. Manage to encourage nonweed species in pasture.
	Poor soil conditions.	Test for soil compaction.
		Determine if soil moisture conditions are too wet or too dry.
		Test soils and address nutrient deficiencies or imbalances.
Pastures are becoming less productive/lower yielding.	Loss of taller-growing, more productive species.	Check grazing management system to make sure plants are not being grazed too short, and that they are allowed enough time to fully regrow after each grazing.
	Plant density is getting lower.	If density is very low, seeding improved pasture species may be needed.
		Correct current and past overgrazing damage.
	Declining soil conditions.	Test for soil compaction.
		Determine if soil moisture conditions are too wet or too dry.
		Test soils and address nutrient deficiencies or imbalances.
	Loss of plant vigor and health.	Correct current and past overgrazing damage.
Pasture quality is declining.	Palatable plants being replaced by less desirable species.	Correct current and past overgrazing damage to productive plant species in pasture.
	Fewer legumes.	Test soils to make sure potassium and boron levels are adequate.
		Check grazing management to make sure short-growing legumes (white clover) are not being shaded out and taller ones (alfalfa) are not being grazed too short.
	Declining soil conditions.	Test for soil compaction.
		Determine if soil moisture conditions are too wet or too dry.
		Test soils and address nutrient deficiencies or imbalances.
	Plants are overmature.	Use shorter regrowth period before grazing.

Table A.3. Soil Problems

Symptom	Cause/Problem	Some Potential Solutions
Yellowing of grass foliage and low productivity.	Possible N deficiency.	Increase amount of legume growing in pasture and/or apply N fertilizer or manure.
Legumes die out or show discoloration.	Possible K or B deficiency.	Test soils, and if K and B are deficient then add according to soil test recommendations.
Low pasture productivity and purple discoloration on some leaves.	Possible P deficiency or low soil temperatures.	Test soils, and if P is deficient then add according to soil test recommendations. If problem is cold soil temperatures, wait a while; symptoms should improve on their own.
Poor pasture productivity.	Possibly nutrient deficiencies.	Test soil so deficiencies can be determined. Then apply soil amendments according to recommendations. If soils are not deficient, then probable cause is poor past management. In this situation, fix grazing management!
Crusty soil surface and poor germination of newly seeded pasture plants.	Lack of litter or plant cover protecting soil surface, resulting in poor water penetration into soil.	Increase plant density, utilize trampling and hoof impact to break up crust, leave higher plant residue in pasture after grazing. Improve soil health and structure.
Soil compaction.	Can be caused by tillage, heavy machinery, or hoof traffic.	Allow longer recovery periods and encourage deeper-rooted pasture plants. Consider chisel plowing or other mechanical methods if biological methods don't work. Improve soil health and structure. Keep livestock off pastures in winter when soils are thawed and saturated.
Mud or standing water.	Wet soils.	Improve drainage to move standing water off pastures. In addition, improving soil structure can improve drainage. If standing water continues to be an issue, additional drainage improvements may be needed.
Erosion.	Insufficient plant density, plant cover, and litter on soil surface.	Use shorter grazing periods and leave more plant residual and litter behind at each grazing. Increase plant density. Minimize or avoid tillage in areas at risk of erosion.
Droughty soils.	Soil has poor water-holding capacity.	Increase soil organic matter content.
Undecomposed manure persists in pasture.	Poor biological activity and decomposition.	Leave more post-grazing residue to encourage soil life. Increase soil organic matter content and surface litter.

Pasture Planning Worksheet

Worksheet Instructions

Line A: Refer to page 117 for a reminder on how to estimate the required dry matter demand for different types of animals. On this worksheet, enter only the amount of dry matter you want to provide from pasture, not the total dry matter demand (unless you plan to supply 100 percent of your herd's needs from pasture).

NOTE FOR ORGANIC FARMERS:

If you are a certified organic farmer and need to calculate % DMI from pasture, use this equation:

$$\% \text{ DMI from pasture} =$$
$$(\text{lbs DM from pasture} \div \text{total DMD}) \times 100$$

Line B: Enter the total number of animals you are planning for.

Line C: Multiply the quantity you entered on line A by the quantity on line B.

Line D: Refer to chapter 13 for instructions on figuring available dry matter per acre.

Line E: Divide the quantity you entered on line C by the quantity from line D. If you have expressed values in terms of acres and end up with an odd value, such as 0.96 acre per day, you can convert to square feet by multiplying the result by 43,560. In the case of this example, that would be 41,818 square feet (0.96 x 43,560). That's equivalent to a paddock that is 200 feet by 209 feet.

Line F: Now that you know how many acres are needed for the herd for one day, you can calculate how many acres will be needed as the length of the regrowth periods change throughout the grazing season. You will need to know the average pasture regrowth periods for your farm or area. If you don't have this information you can find it through your local extension service, NRCS, grazing consultant, or grazing network.

In the chart for line F, enter the name of each month, the number of days of regrowth needed in each month, and the acres needed per day (from line E) to supply the necessary amount of dry matter. Don't forget to add one paddock to account for the one they are grazing!

Which month is the most acreage required? Do you have enough land?

Pasture Planning Worksheet

Use this worksheet to calculate how large a paddock is required for one grazing group for a single day, and then to calculate the total acres needed to sustain that grazing group throughout the grazing season.

Line A: Pounds dry matter required from pasture per animal per day   _____

Line B: Number of animals in the group   _____

Line C: Total pounds dry matter from pasture needed per day for the group

Line A _____ × Line B _____ = _____

Line D: Pounds of dry matter *available* per acre   _____

Line E: Total number of acres needed per day

Line C _____ ÷ Line D _____ = _____

Line F: Number of acres needed each month throughout the grazing season

Month	Regrowth Period	× Acres/Day (plus 1 paddock)		
_____	_____	× _____	=	_____
_____	_____	× _____	=	_____
_____	_____	× _____	=	_____
_____	_____	× _____	=	_____
_____	_____	× _____	=	_____
_____	_____	× _____	=	_____
_____	_____	× _____	=	_____
_____	_____	× _____	=	_____

Pasture Monitoring Worksheet

Directions: Walk through each pasture area on your farm and use the chart below to assess its quality in each category. Rank each pasture area as poor, fair, good, or very good in each category. Include comments and descriptions of what you observe to help you track changes in your pastures over time.

Category	Score: Poor, Fair, Good, Very Good	Notes and Comments
Plant diversity: How many different species of plants are in the pasture? A poor pasture will have few legumes and a small number of grass species. A good-quality pasture will contain more perennial species as well as two or more legume species and many grass species.	_____	_____
Plant density: How much vegetative cover is in the pasture? Poor density will have soil visible between the plants, while a good-quality pasture will be higher-density and will have more complete cover.	_____	_____
Palatability of plants: How many of the plants in the pasture are what the animals want to eat? A good-quality pasture will contain mostly palatable species at the stage of maturity livestock prefer to graze. A poor-quality pasture will contain nonpalatable weed species and other plants that livestock will not eat.	_____	_____
Plant growth rate: How vigorously do the plants in the pasture grow over the whole growing season? Healthy productive pasture should grow well in the spring, summer, and fall without extreme changes in productivity as long as soil moisture and temperatures allow.	_____	_____
Soil health: How fast do plant residue and manure decompose? Is there ample organic matter? Are there signs of fertility deficiencies or imbalances? Are soils compacted, too droughty, or too wet?	_____	_____

Third-Party Certification Programs

Farmers who are marketing a product that requires some type of third-party certification may want to have their farms and products certified organic, or may seek certification through one of the grassfed certification programs. Each of these third-party certification programs has different standards relating to pasture management and grazing. Information on all the details of these standards is beyond the scope of this book, and the standards may be changed over time. It is important to consult an up-to-date, complete version of the standard before applying for any type of certification program.

Organic Certification Pasture Requirement

Organic standards in the United States are set by the US Department of Agriculture (USDA) Agricultural Marketing Service (AMS) National Organic Program (NOP). These standards include requirements for organic farms and food manufacturers that want to sell or label their products as organic. Within these comprehensive standards are specific requirements for farms that want to certify milk or meat from ruminants as organic. For farms in other countries, the organic standards will be different, so some additional research will be needed.

In the United States, there are many third party certifiers that offer organic certification. A full list of accredited organic certification agencies can be found on the USDA AMS website (www.ams.usda.gov/rules -regulations/organic).

The organic standards for livestock include requirements on feed, housing, health care, record keeping, livestock purchases, and crop management. Within the standards there are very specific requirements regarding pasture for ruminants. A few key points of these requirements include:

- All livestock feed (including all pastures) must be certified organic.

- Each group six months or older on the farm must get at least 30 percent of their dry matter intake needs from pasture during the grazing season. This is an average over the whole grazing season.
- The grazing season cannot be any shorter than 120 days, but is expected to be longer than that in many regions of the United States.
- Farms are required to keep records of what stored rations are fed to each group so that certifiers can verify how much dry matter in the ration comes from pasture versus supplemental feed.
- Pasture must be managed as a crop, and a pasture plan is required.
- There is a list of exemptions for when grazing is not required. The list includes things such as treating sick animals, drying off lactating animals, calving, breeding, and milking.

These are just a few of the key points in the organic standards relating to pasture; a full version can be found online. Once a farm chooses a certifier, they will provide additional information on required record keeping, as well as the application process and costs. Costs of organic certification vary from certifier to certifier, but the USDA also has a program that reimburses certification costs up to $750 per year per scope that is certified.

Since the organic standards require a relatively small amount of dry matter intake from pasture during the grazing season (only 30 percent), there is considerable interest in grassfed certification so that farms can demonstrate they are providing more feed from pasture. There are several grassfed certification programs and standards at this time; let's look at some of the details now.

Grassfed Certification

Several grassfed certification programs are available in the United States. Each has differing requirements and costs. Due to increasing consumer and market interest in grassfed products and certification, these standards and certification options are changing rapidly — be sure to check the current standards and current lists of certifiers. At the time of publication, the information below is accurate, but it may change quickly.

GRASS FED — USDA

This program was introduced in 2007 through the USDA's Livestock Process Verified Program. It is a voluntary standard that applies only to meat products from ruminants marketed with a USDA-verified grassfed claim. Milk and animal fiber products are not included in the scope of this standard.

This certification is a "desk audit" based on required forms and does not include an on-farm inspection. The producer must be able to verify that the grassfed marketing claim standard requirements are being met through a detailed documented quality management system.

All label claims, including the ones verified by a USDA Process Verified Program, must be approved by USDA's Food Safety Inspection Service (FSIS) Label Protection and Delivery Division (LPDD).

Summary of the standards: Animals cannot be fed grain or grain by-products and must have continuous access to pasture during the growing season. Crops normally harvested for grain are eligible feed only if they are grazed or harvested in the vegetative state (pre-grain). Animals must be managed to grassfed standards from birth to slaughter.

The costs vary for this program, with estimates ranging from $500 to $2,000 annually based on the size of the operation and complexity of the audit.
www.ams.usda.gov

GRASS FED FOR SMALL AND VERY SMALL PRODUCERS — USDA

This program was introduced in 2014. Also developed by the USDA, this one is only available to small producers — defined as producers marketing 49 cattle or less each year and lambs produced from 99 ewes or less. This is also a desk audit certification based on required forms. However, an on-site audit may be required depending on how complex the operation is. Producers who are

certified under this program receive a certificate that will allow them to market live cattle and sheep as USDA-certified grassfed.

The standards are similar to Grass Fed — USDA certification with additional requirements of operation size. The cost is significantly lower — most recently listed at $108 for two years of certification.
www.ams.usda.gov

AMERICAN GRASSFED BY AMERICAN GRASSFED ASSOCIATION (AGA)

This program was introduced in 2009 by a group of producers, food service industry personnel, and consumer interest representatives who established AGA. Standards apply to all farms and ranches approved by AGA for the production of ruminant animals and products.

This program requires that the standards be verified by an independent, third-party, on-farm yearly audit. Once certified, the farm can use the AGA's logo, trademark, service mark, and/or design mark.

These standards state that pasture is required and animals must be maintained at all times in areas with at least 75 percent forage cover. During periods of low forage quality due to the season and/or inclement weather, producers can also feed hay, haylage, balage, silage without grain, forage products, crop residue without grain, and other approved roughage sources. Feeding of supplements must be within 1 percent of lifetime total dry matter intake for the animal. The standards prohibit feeding grain products in any form; antibiotics, hormones, and organophosphates; milk replacer that contains any prohibited health care products; and molasses fed with a target daily intake greater than 3 pounds per head. Animals must be born in the United States and accompanied by records that verify their identity and grassfed management from birth to slaughter.

The cost for an approved AGA Producer is $100 per year for membership dues and a $100 licensing fee. In addition, Approved Producer members pay a per-head fee — currently $1 per head for large ruminants (beef or bison) or per head milked, and/or 25 cents for small ruminants (goat and sheep) for every animal harvested for the Approved Producer program.
www.americangrassfed.org

AWA Grassfed — Animal Welfare Approved

The AWA Grassfed standards were introduced in 2014 and verify that farms are in compliance with AWA beef, meat and dairy sheep, meat and dairy goat, and/or bison standards (not applicable for dairy cattle at this time).

This program requires an application and annual on-site inspection. It also requires the use of AWA-approved slaughter facilities. Once certified, farms can use the AWA-certified grassfed and AWA labels and other promotional materials.

The standard requires that with the exception of milk consumed prior to weaning, the diet of grassfed animals must solely be derived from grass and forage throughout their lives. Grassfed animals may be supplemented with hay, haylage, balage, silage, crop residue (straw) without grain, and other natural sources of roughage while on pasture. Mineral and vitamin supplements must not include any prohibited ingredients as outlined in the grassfed and AWA species-specific standards. Some examples of prohibited ingredients under the AWA standards include animal by-products, fish meal, subtherapeutic antibiotics, and organophosphates. This program has no cost to the farmer. Grants are available to help farmers improve animal welfare on the farm.

www.animalwelfareapproved.org

100% Grassfed — Pennsylvania Certified Organic or NOFA NY Certified

Introduced in 2014 by Pennsylvania Certified Organic (PCO), this program was designed for dairy farmers but also covers meat. NOFA NY Certified Organic LLC now offers a parallel program.

This program requires an application, which is an add-on to the farm's organic certification application. It also includes on-site inspection, which can be done in combination with the annual organic inspection. Once certified, the farm can use the PCO or NOFA NY 100% Grass Fed seal on certified product labeling and marketing materials.

To achieve this certification, farms must already meet the USDA organic standards. There is a required period of transition in which dairy cows must be on a grassfed (no grain) diet prior to certification. Following the transition, animals must remain under grassfed management. Molasses is allowed, grain is not allowed, and there are some limitations on some other supplemental feeds.

For cost estimates, refer to the NOFA NY Certification or PCO Fee Schedule on their websites.

www.paorganic.org/grassfed
www.nofany.org/100%25grassfedcertification

—APPENDIX E—

Resources

EXTENDING THE GRAZING SEASON

Gerrish, Jim. *Kick the Hay Habit: A Practical Guide to Year-Around Grazing*. Green Park Press, 2010.

Chapter 6 in *Forage Utilization for Pasture-Based Livestock Production*[1] includes a method to calculate the value of extending the grazing season.

FORAGE QUALITY AND TESTING

Henning, Jimmy C., Garry D. Lacefield, and Donna Amaral-Phillips. "Interpreting Forage Quality Reports." ID-101. University of Kentucky Cooperative Extension, 1991. Available online at http://www2.ca.uky.edu/agc/pubs/id/id101 /id101.htm.

Flack, Sarah, and Karen Hoffman. 2013. "Managing Dairy Nutrition for the Organic Herd: Forage Testing and Interpreting Lab Analysis." Available online at http://articles.extension.org/pages /68573/managing-dairy-nutrition-for-the -organic-herd:-forage-testing-and-interpreting -lab-analyses.

Rasby, Rick J., Paul J. Kononoff, and Bruce E. Anderson. "Understanding and Using a Feed Analysis Report." G1892. University of Nebraska, Lincoln Extension, 2008. Available online at http://extensionpublications .unl.edu/assets/pdf/g1892.pdf.

GRAZING BEHAVIOR AND PLANT– HERBIVORE INTERACTIONS

Emmick, Darrell L., and Frederick D. Provenza. "Animal Ecology and Foraging Behavior." Chapter 1 in *Animal Production Systems for Pasture-Based Livestock Production*, edited by Edward B. Rayburn. Natural Resource, Agriculture, and Engineering Service (NRAES), 2007.

Provenza, Frederick D. *Foraging Behavior: Managing to Survive in a World of Change*. USDA NRCS, 2003.

GRAZING MANAGEMENT METHODS

Rayburn, Edward B., editor. *Forage Utilization for Pasture-Based Livestock Production*. Natural Resource, Agriculture, and Engineering Service (NRAES), 2007.

Gerrish, Jim. *Management Intensive Grazing: The Grassroots of Grass Farming*. Green Park Press, 2004.

———. Kick the Hay Habit: *A Practical Guide to Year-Around Grazing*. Green Park Press, 2010.

Jim Gerrish also offers grazing schools, listed on his website: www.americangrazinglands.com/events.

Hoffman Sullivan, Karen, Robert J. DeClue, and Darrell L. Emmick. *Prescribed Grazing and Feeding Management of Lactating Dairy Cows*. NYS Grazing Lands Conservation Initiative, USDA Natural Resources Conservation Service, 2000.

Emmick, Darrell L. *Managing Pasture as a Crop: A Guide to Good Grazing*. University of Vermont Extension, n.d. Available online at http://anr.ext.wvu.edu/r /download/195210.

For information on the grazing wedge pasture planning and record-keeping system, see http://grazingwedge .missouri.edu.

Savory, Allan. *Holistic Management: A New Framework for Decision Making*. Island Press, 1998.

Salatin, Joel. *Salad Bar Beef*. Polyface, 1996.

Murphy, Bill M. *Greener Pastures on Your Side of the Fence*. Arriba Publishing, 1991.

INFRASTRUCTURE: WATER SYSTEM AND FENCE DESIGN

Gerrish, Jim. *How to Management Guides: Center Pivot Grazing Guide, Electric Fence Basics,* and Stock Water Development Guide. Available online at www .americangrazinglands.com/wp-content/uploads /2015/04/Pasture-Management-Tools-Online.pdf.

Rayburn, Edward B., editor. *Animal Production Systems for Pasture-Based Livestock Production*. Natural Resource, Agriculture, and Engineering Service (NRAES), 2007.

Missouri USDA NRCS. 2005. "Electric Fencing For Serious Graziers." Available online at www.nrcs .usda.gov/Internet/FSE_DOCUMENTS/nrcs144p2 _010636.pdf.

"Forage Crop Irrigation Systems and Economics." G1697. University of Missouri Extension, 2013. Available online at http://extension.missouri.edu/p/G1697.

Byelich, Boyd, Jennifer Cook, and Chayla Rowley. "Small Acreage Irrigation Guide." USDA NRCS and Colorado State University Extension (2013): Available online at www.ext.colostate.edu/sam/sam -irr-guide.pdf.

LAND AND GRAZING PLANNING

Acreage and distance measurements are available (free) at the website measurelotsize.com, as well as (not free) the app GeoMeasure.

GoGraze is an app for grazing planning.

See http://websoilsurvey.sc.egov.usda.gov for soil maps for your farm.

To learn the soil types in your location, see the app SoilWeb.

http://holisticmanagement.org/free-downloads is a source of grazing planning charts.

Grazing wedge is available at the website http://grazing wedge.missouri.edu.

PASTURE DRY MATTER MEASUREMENTS AND TOOLS

There are several online resources with instructions and video demonstrations on how to measure pasture dry matter. Some of these can be found on the author's web-site: www.sarahflackconsulting.com. You can also check: www.noble.org/ag/pasture/grazingstick www.sarahflackconsulting.com/publications-and -video/videos.

For a list of where to get a grazing stick, see http:// articles.extension.org/pages/28873/finding-a -pasture-stick-in-your-area-for-your-organic-dairy -farm#.Vkx_c2SrTak.

PASTURE PLANT SPECIES AND IDENTIFICATION

Abaye, A. Ozzie. *Identification and Adaptation: Common Grasses, Legumes and Non-Leguminous Forbs of the Eastern United States*. Virginia Tech, 2010.

Rayburn, Edward B., editor. *Forage Production for Pasture-Based Livestock Production*. Natural Resource, Agriculture, and Engineering Service (NRAES), 2007.

Ernst Conservation Seeds, Inc. *Native Warm Season Grasses for High Quality Biomass Forage, Including Livestock Bedding and Mushroom Compost*. Available online at www.ernstseed.com/files/documents /native_warm.pdf.

Undersander, Dan, Michael Casler, and Dennis Cosgrove. *Identifying Pasture Grasses*. University of Wisconsin, 1996. Available online at http://learning store.uwex.edu/assets/pdfs/a3637.pdf.

Undersander, Dan, and Dennis Cosgrove. *Identifying Pasture Legumes*. University of Wisconsin, 2003. Available online at http://learningstore.uwex.edu /assets/pdfs/a3787.pdf.

RUMINANT NUTRITION, DRY MATTER DEMAND, BODY CONDITION, AND HEALTH ON PASTURE

Neary, Michael, and Ann Yager. "Body Condition Scoring in Farm Animals." AS-550-W. Purdue University Department of Animal Sciences, 2002. Available online at www.extension.purdue.edu /extmedia/as/as-550-w.pdf.

National Research Council. *Nutrient Requirements of Dairy Cattle*, 7th revised ed. National Academy Press, 2001.

National Organic Program Handbook: Guidance and Instructions for Accredited Certifying Agents and Certified Operations. NOP 1100. USDA AMS NOP (National Organic Program), 2011. DMD tables available online at www.ams.usda.gov/sites/default /files/media/Program%20Handbk_TOC.pdf.

Eversole, Dan E., Browne, Milyssa F., Hall, John B., and Dietz, Richard E. "Body Condition Scoring Beef Cows." 400-795. Virginia Cooperative Extension, 2009. Available online at https://pubs.ext.vt.edu /400/400-795/400-795.html.

Kellogg, Wayne. "Body Condition Scoring with Dairy Cattle." FSA4008. University of Arkansas Division of Agriculture, n.d. Available online at www.uaex.edu /publications/PDF/FSA-4008.pdf.

Greiner, Scott P. "Ewe Body Condition Scoring." n.d. Available online at www.apsc.vt.edu/extension /sheep/programs/shepherds-symposium012/12 _symposium_greiner_bcs.pdf.

Rinehart, Lee, and Ann Baier. *Pasture for Organic Ruminant Livestock.* NCAT May 2011. Available online at https://attra.ncat.org/attra-pub/summaries /summary.php?pub=360.

For parasite information on both FAMACHA and fecal egg testing, see www.wormx.info.

For information on parasite life cycles, management, and natural products see http://eap.mcgill.ca /agrobio/ab370-04e.htm.

SEEDING

Rayburn, Edward B., editor. *Forage Production for Pasture-Based Livestock Production.* Natural Resource, Agriculture, and Engineering Service (NRAES), 2007.

Abaye, A. Ozzie *Identification and Adaptation: Common Grasses, Legumes and Non-Leguminous Forbs of the Eastern United States.* Virginia Tech, 2010.

Bosworth, Sid. *Description and Seeding Rates for Forage Plants Grown in Vermont.* University of Vermont Extension Service, 2012. Available online at www. uvm.edu/pss/vtcrops/articles/VT_Forage _Description_and_Seeding_Rate.pdf.

SOIL FERTILITY AND HEALTH

Sullivan, Preston. *Sustainable Soil Management.* ATTRA (Appropriate Technology Transfer for Rural Areas), 2001.

Brunetti, Jerry. *The Farm as Ecosystem: Tapping Nature's Reservoir — Biology, Geology, Diversity.* Acres USA, 2014.

SOIL TESTING LABS

Most state extension services or universities have a local soil test lab. Additional labs include:

A&L Laboratories, www.al-labs-eastern.com.

Logan Labs, www.loganlabs.com.

LABS AND RESOURCES FOR MORE COMPREHENSIVE SOIL HEALTH ASSESSMENT AND TESTING

Cornell University Soil Health Testing Services, http:// soilhealth.cals.cornell.edu/testing-services.

NRCS Soil Health Assessment, www.nrcs.usda .gov/wps/portal/nrcs/main/soils/health/assessment.

NOTES

INTRODUCTION

1. Sarah Flack, "The Relationship of Light and Plant Response in a Complex Pasture Sward with Particular Emphasis on White Clover" (master's thesis, University of Vermont, 1994).

CHAPTER 1

1. Jo Robinson, *Pasture Perfect* (Vashon Island Press, 2011), 30–34.
2. Ibid., 37–39.
3. Kate Clancy, *Greener Pastures* (Union of Concerned Scientists, 2006), 14.

CHAPTER 2

1. André Voisin, *Grass Productivity* (Island Press, 1959).
2. Ibid., 7.
3. Jim Gerrish, *Management Intensive Grazing: The Grassroots of Grass Farming* (Green Park Press, 2004).
4. Allan Savory, *Holistic Management: A New Framework for Decision Making* (Island Press, 1998).

CHAPTER 3

1. Flack, "The Relationship of Light and Plant Response."
2. Robert L. Kallenback, "Dairy Grazing: Growth of Pasture Plants," University of Missouri Extension Dairy Grazing Publication Series M182, February 2012.
3. Charles Walters and Gearld Fry, *Reproduction and Animal Health* (Acres USA, 2003).
4. John Hendrickson and Bret Olson, "Understanding Plant Response to Grazing," in *Targeted Grazing: A Natural Approach to Vegetation Management and Landscape Enhancement* ed. Karen Launchbaugh, (American Sheep Company Association, 2006), www.sheepusa.org/targetedgrazing, 37.
5. Ibid.
6. Kenneth J. Moore et al., "Describing and Quantifying Growth Stages of Perennial Forage Grasses," *Agronomy Journal* 83 (1991): 1073–77.

7. Kallenback, "Dairy Grazing."
8. Flack, "The Relationship of Light and Plant Response."

CHAPTER 4

1. Newman Turner, *Fertility Pastures* (Acres USA, 2009, originally published in 1955 by Faber and Faber), 3.
2. Jerry Brunetti, *The Farm as Ecosystem: Tapping Nature's Reservoir — Biology, Geology, Diversity* (Acres USA, 2014), 1.
3. Ibid, 82.
4. Fred Magdoff and Harold van Es, *Building Soils for Better Crops* (Sustainable Agriculture Network, 2000), 7.
5. Sir Albert Howard, Introduction to J. Rodale, "Pay Dirt" (Devin Adair Company, 1945), 4.
6. Preston Sullivan, *Sustainable Soil Management* (ATTRA, 2001), 10.
7. Dr. Hubert Karreman, *Treating Dairy Cows Naturally* (Acres USA, 2007), 5.
8. Bill M. Murphy, *Greener Pastures on Your Side of the Fence* (Arriba Publishing, 1991).
9. Edward B. Rayburn, editor, *Forage Production for Pasture-Based Livestock Production* (Natural Resource, Agriculture, and Engineering Service, 2007).
10. Ibid.

CHAPTER 5

1. James Anderson, *Essays Relating to Agriculture and Rural Affairs* (W. Creech, 1777).
2. Ibid., Volume 2, 296
3. Voisin, *Grass Productivity*, 333.
4. Ibid.
5. Sid Bosworth, *Description and Seeding Rates for Forage Plants Grown in Vermont*, University of Vermont Extension, 2012.
6. Jim Gerrish, *Kick the Hay Habit: A Practical Guide to Year-Around Grazing* (Green Park Press, 2010).

7. Edward B. Rayburn, editor, *Forage Utilization for Pasture-Based Livestock Production* (Natural Resource, Agriculture, and Engineering Service, 2007), 82.

Chapter 6

1. David B. Hannaway and Marc Cool, "Meadow Fescue (*Festuca pratensis* Huds.)," 2004, http://forages.oregonstate.edu/php/fact_sheet_print_grass.php?SpecID=75.

Chapter 7

1. Jane A. Parish and J. Daniel Rivera, "Understanding the Ruminant Animal Digestive System," Mississippi State Extension Service, 2009.
2. Darrell L. Emmick and Frederick D. Provenza, "Animal Ecology and Foraging Behavior," in *Animal Production Systems for Pasture Based Livestock Production*, ed. Edward B. Rayburn (Natural Resource, Agriculture, and Engineering Service, 2007).
3. Bill Murphy, *Greener Pastures on Your Side of the Fence*, 15.
4. Emmick and Provenza, "Animal Ecology and Foraging Behavior."
5. Gerrish, *Management Intensive Grazing*.

Chapter 8

1. Rayburn (ed.), *Animal Production Systems for Pasture-Based Livestock Production*.
2. Rick J. Rasby, Paul J. Kononoff, and Bruce E. Anderson. "Understanding and Using a Feed Analysis Report," University of Nebraska–Lincoln Extension G1892, 2008, http://extensionpublications.unl.edu/assets/pdf/g1892.pdf.
3. Karen Hoffman, "Managing Dairy Nutrition," in *The Organic Dairy Handbook*, ed. Katherine Mendenhall (Northeast Organic Farming Association of New York, 2009).
4. Sarah Flack and Karen Hoffman, "Managing Dairy Nutrition for the Organic Herd: Forage Testing and Interpreting Lab Analyses," http://articles.extension.org/pages/68573/managing-dairy-nutrition-for-the-organic-herd:-forage-testing-and-interpreting-lab-analyses.

Chapter 9

1. Voisin, *Grass Productivity*, 67.
2. Hoffman, "Managing Dairy Nutrition."

Chapter 10

1. Frederick D. Provenza, "Foraging Behavior: Managing to Survive in a World of Change," Utah State University, 2003, http://extension.usu.edu/behave.
2. Ibid.

Chapter 12

1. Rayburn (ed.), *Animal Production Systems for Pasture-Based Livestock Production*.
2. Provenza, "Foraging Behavior."
3. Charles L. Rhykerd and Keitit D. Johnson, "Minimizing the Prussic Acid Poisoning Hazard in Forages," Purdue University Cooperative Extension Service, 2007, www.agry.purdue.edu/ext/forages/publications/ay196.htm.

Chapter 13

1. Stephen K. Barnhart, "Estimating Available Pasture Forage," Iowa State University Extension, 2009.

Chapter 14

1. Rayburn (ed.), *Forage Utilization for Pasture-Based Livestock Production*.

Chapter 15

1. Savory, *Holistic Management*.
2. Heather Darby and Sarah Flack, "Fine Tuning Your Grazing System," www.sarahflackconsulting.com.
3. Savory, *Holistic Management*.
4. USDA Natural Resources Conservation Service, Pasture Condition Score Sheet, www.nrcs.usda.gov/Internet/FSE_DOCUMENTS/stelprdb1044243.pdf.
5. Holistic Management International, http://holisticmanagement.org.

Appendix E

1. Rayburn (ed.), *Forage Utilization for Pasture-Based Livestock Production*, 2007.

GLOSSARY

Acid detergent fiber (ADF): The fibrous portion of plant cell walls, which are made up of cellulose and lignin. These are the least digestible portions of roughage, so as ADF levels increase, digestible energy levels in the forage decrease.

Carbon sequestration: The removal of carbon from the atmosphere and storage of the carbon in plants, soil, or the ocean through processes that may be physical or biological (such as photosynthesis).

Crown: The basal zone of the shoot in perennial grasses where the overwintering tissues and buds are found, and which will produce new growth following dormancy.

Crude protein (CP): The percentage of forage that supplies nitrogen or amino acids (protein).

Dry matter or pasture dry matter: The amount of forage after all the water has been excluded. This allows us to better understand how much actual nutrition there is in the plant material.

Dry matter demand (DMD): The amount of feed, on a dry matter basis, that a specific class of animal is predicted to eat.

Dry matter intake (DMI): The amount of feed an animal eats excluding the water content.

Erosion: The wearing away of the land surface by running water, wind, ice, or other geological agents, including such processes as gravitational creep. Also, movement of soil or rock fragments by water, wind, ice, or gravity.

Forb: A flowering plant in pasture that is not a legume or a grass. Forbs may include forage chicory, dandelion, and plantain.

Grazer: An animal that eats pasture.

Grazier: A farmer who manages grazing animals.

Growing point: An area of a plant where new growth occurs. In pasture plants this includes the tips of roots as well as apical growing points at the tips of stolons. There are also growing points in buds, which may be found in the crown of perennial grasses, and adventitious buds from which tillers/stems emerge elsewhere on the plant. Grasses also have growing points known as intercalary meristems at the collar or base of the leaf blade, which allows continued leaf growth from its base.

Gully: A type of erosion in which a channel is formed because of concentrated flow of surface or stormwater runoff over unprotected erodible land.

Herbivore: An animal that eats plants.

Holistic Planned Grazing (HPG): A grazing system developed by Allan Savory as part of the more comprehensive Holistic Management system.

Leaf area index (LAI): A measure of how much leaf area there is in the plant canopy. It is defined as the amount of leaf surface per area of ground. This determines how much sunlight the pasture can capture and use for photosynthesis.

Management intensive Grazing (MiG): A term coined by Jim Gerrish to describe well-managed grazing systems in which animal nutritional needs are balanced with forage supply. This emphasis on good management is why the acronym has a capital M, but a lowercase i.

Mass wasting: A form of erosion that involves movement of large volumes of earth material downslope.

Moisture content: The percentage of water in a forage sample when it was tested.

Neutral detergent fiber (NDF): A measure of fiber in a forage. It measures the fiber in the cell walls and includes the same fiber measured by ADF with the addition of hemicellulose. As NDF percent increases in a forage, dry matter intake by the livestock generally decreases.

Paddock: A small fenced area used for grazing.

Palatability: The degree to which food is agreeable to the palate or taste.

Post-ingestive feedback: Feedback from the gut to the brain allowing animals to sense the nutritional or toxicological effects of a food. This positive or negative feedback lets them adjust their preference for a type of food and choose to eat more or less of it.

Prescribed Grazing: A term used by NRCS to describe systems in which management of vegetation is done correctly, using grazing or browsing livestock.

Recovery period: The amount of time pasture plants are allowed to regrow or "rest" after each grazing.

Relative feed value (RFV): A calculated value of overall digestibility and forage intake potential.

Residual: Living plant material left behind after grazing.

Residue: The litter, or dead plant material, left on the soil surface after grazing.

Rhizome: A horizontal underground stem or root.

Rill: A form of erosion that's similar to a gully, but only a few inches deep. Often rills are caused by an increase in surface water flow when soil is not protected by vegetation.

Rotational Grazing: A simple system in which livestock are rotated through a number of paddocks, often according a fixed schedule rather than on observation of plant growth rates.

Ruminants: Cows, sheep, goats, and other animals that have a rumen as part of a four-stomach digestive system.

Sheet erosion: A form of erosion in which the soil surface is removed by surface runoff of a fairly uniform layer.

Sloughing: A form of erosion in which a mass of soil moves down a slope, similar to a landslide; also known as slumping.

Stocking density: The number of cows per acre or pounds of animals per acre during just the short period of time that they are in an individual paddock. This is different from the stocking rate, which refers to the total number of animals on the entire farm.

Stocking rate: The long-term carrying capacity of the farm or the total number of animals on the entire farm.

Stockpiled pasture: Forage that is allowed to grow and accumulate for use at a later time or during a period of forage or pasture deficit.

Stolon: A stem that grows horizontally over the surface of the soil. It has roots at points along this stem, which may also be points where new plants can form.

Tiller: The stem produced by a grass plant. Each new shoot that grows after the first one emerges from the seed is a new tiller.

Total digestible nutrients (TDN): A calculated value of overall digestibility or energy value of a forage.

Voisin grazing: A grazing system named after scientist and farmer André Voisin, though he actually referred to it as "rational" grazing. This system, which many of the modern well-managed grazing systems evolved from, is based on three fundamental guidelines for grazing: allow plants sufficient time to regrow after each grazing; a short period of occupation in each paddock of three days or less; and allow livestock with the highest nutritional requirements to harvest the best pasture.

Wind erosion: Erosion caused when soil particles are removed by wind, usually from dry areas of soil that are not protected by vegetation.

INDEX

ABOUT THE AUTHOR

Melanie Ross

Sarah Flack is the author of *Organic Dairy Production* and is a nationally known speaker and consultant on grazing and organic livestock. She grew up on a Vermont family farm that used management-intensive grazing and mob stocking. She later studied Holistic Planned Grazing and did her graduate studies on pasture management at the University of Vermont. She has written extensively about grass farming and has taught workshops and consulted with farms both in the United States and internationally. Sarah's approach in her consulting, writing, and teaching is to help create more farms with successful grass-based management systems, empowering farmers to create positive change for their pastures, soils, livestock, finances, and farm-family quality of life. When she is not traveling, Sarah lives in northern Vermont on her solar-powered small farm.

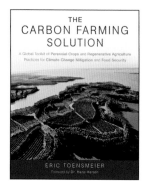
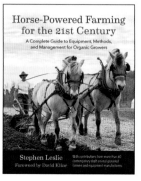
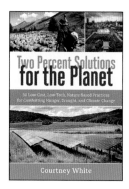
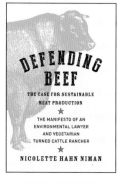